Wind Energy:
Societal and Human Impacts

Wind Energy: Societal and Human Impacts presents the theoretical basis for the various impacts of wind turbines on humans. These impacts include noise (in the audible, low-frequency and infrasound ranges), visual effects (shadow flicker and light reflections), electromagnetic fields, vibrations and oscillations from wind turbines, as well as mechanical impacts and physical risk related to turbine collapse or turbine component failure, throwing of ice or blade parts, fires, etc. Further, this book examines the legal framework, policy approaches and investment processes in the wind energy sector. It addresses the issue of siting and minimum distances between turbines and residential buildings and proposes modifications to legislation, as well as guidelines and recommendations related to environmental impact reports for wind farms, assessment of such reports, monitoring of such impacts and the performance of a post-implementation analysis.

Prof. Piotr Kacejko, Ph.D., is a graduate of the Faculty of Electrical Engineering of Lublin University of Technology. He specializes in the analysis of electric power systems. He is co-author of computer programs used in the power sector and author of more than 200 articles, papers and research reports. His work carried out over the past 15 years has concerned the determination of connection possibilities, the operation of RES units in the power system and the study of stability in system development planning. He is an expert appointed by the Energy Regulatory Office. He is also a member of the Committee on Electrical Engineering of the Polish Academy of Sciences, and a member of the Association of Polish Electrical Engineers and the American IEEE. He served as Rector of Lublin University of Technology from 2012 to 2020.

Jacek Szulczyk, Ph.D., is a doctor of technical sciences with specialization in vibroacoustics and a graduate of the Poznan University of Technology. For almost 10 years he cooperated with the Department of Vibroacoustics and Biodynamics of Systems, first as a doctoral student, then as an assistant. He currently works as Chief Acoustician at the Acoustics and Environmental Laboratory of EKO-POMIAR. He is a member of the National Commission for Environmental Impact Assessment. For more than 10 years he has been carrying out acoustic tasks related to the design and measurement of various noise sources, including wind turbines.

Adam Zagubień, Ph.D., is a researcher and teacher at the Koszalin University of Technology. He has 20 years of measurement practice in the field of acoustics and vibration. He is a contractor and manager in several research projects on vibration and noise analysis. He has authored and co-authored more than 30 scientific articles, and published in foreign and domestic journals. He is a reviewer for several foreign journals. He is also an author and co-author of more than 300 expert reports in the field of environmental protection against noise and vibration, for industrial plants, roads, ports, ships, passenger car ferries and wind farms.

Wind Energy:
Societal and Human Impacts

Piotr Kacejko, Jacek Szulczyk, and Adam Zagubień

CRC Press
Taylor & Francis Group
Boca Raton London New York

CRC Press is an imprint of the
Taylor & Francis Group, an **informa** business

Designed cover image: Adam Zagubień

First edition published 2024
by CRC Press
2385 NW Executive Center Drive, Suite 320, Boca Raton FL 33431

and by CRC Press
4 Park Square, Milton Park, Abingdon, Oxon, OX14 4RN

CRC Press is an imprint of Taylor & Francis Group, LLC

© 2024 Taylor & Francis Group, LLC

ISBN: 978-1-032-59878-9 (hbk)
ISBN: 978-1-032-59880-2 (pbk)
ISBN: 978-1-003-45671-1 (ebk)

DOI: 10.1201/9781003456711

Typeset in Times
by codeMantra

Contents

Acknowledgment

The authors would like to thank Professor Andrzej W. Jasiński and Dr. Katarzyna Matuszczak for their help in collecting material for this book and for their valuable comments.

Introduction

The Role of RES in Energy Transition with Particular Emphasis on Wind Power

An analysis of the current crises occurring worldwide indicates the growing need for a global energy transition. Centralized energy systems are heavily dependent on fossil fuels. Increasing prices of major primary energy carriers have a decisive impact on the stability and efficiency of economic systems. The war currently waged in Ukraine clearly illustrates this dependence. In turn, the effects of climate change caused by human economic activity are becoming increasingly acute. According to the reports by the Intergovernmental Panel on Climate Change (IPCC), between 3.3 and 3.6 billion people already live in the areas threatened by severe climate change (IPCC 2022a, 2022b).

The realization of the international community's plan to reduce temperature relative to the pre-industrial period by 1.5°C (and even by 2°C) by the mid-21st century depends on the intensity and scope of the aforementioned energy transition. Its acceleration will also be important for long-term energy security, price stability and energy flexibility. It is estimated that about 80% of the global population lives in countries that are net energy importers (World Energy Transitions Outlook 2022). Given the wealth of renewable energy potential that has yet to be tapped, this percentage could be drastically reduced. This would make these countries less dependent on energy imports through diversified supplies and, to some extent, help make economies impervious to significant fluctuations in fossil fuel prices. The use of RES also means new jobs, lesser poverty as well as the development of a global and climate-safe economy. With each passing day, the cost of inaction in the energy transition increasingly outstrips the cost of action.

The 1.5°C pathway, set by IRENA (International Renewable Energy Agency) as a concept incorporating the conclusions drawn during the 2015 Paris Conference, considers electrification and energy efficiency as key drivers of the energy transition made possible by renewable energy sources, hydrogen and sustainable biomass. This pathway requires a massive change in the way societies generate and consume energy, but could reduce annual CO_2 emissions by nearly 37 gigatons by 2050 (IRENA 2021a).

In general, this reduction can be achieved through:

DOI: 10.1201/9781003456711-1

- increase in the production and direct use of renewable energy;
- improvement of energy efficiency;
- electrification of end-user sectors (e.g., electric vehicles and heat pumps);
- production and use of green hydrogen and its derivatives;
- use of bioenergy combined with carbon capture and storage;
- use of sequestered and stored carbon dioxide at the end user (IRENA 2022).

In order to meet the goals of the Paris Agreement, the CO_2 emissions related to energy generation would have to be reduced by about 3.5% per year from now until 2050, assuming that these reductions continue beyond that period. A shift to increasingly electrified forms of transportation and heat generation, combined with an increase in renewable energy generation, would provide about 60% of the reductions in energy-related CO_2 emissions planned by 2050. If additional reductions resulting from the direct use of renewable energy sources are included, the share rises to 75%. Adding energy efficiency further increases this share to more than 90% (IRENA 2019a).

The energy transition would also increase global gross domestic product (GDP) by 2.5% and total employment by 0.2% in 2050. In addition, it would bring greater social and environmental benefits. The health, subsidy and climate savings combined would be worth a total of $160 trillion over 30 years. Thus, every dollar spent on transforming the global energy system provides a return of at least $3, and potentially more than $7, depending on the valuation of externalities (IRENA 2019a).

Realizing the energy transition at the required pace and scale would require almost complete decarbonization of the electricity generation sector by 2050. This would lead to an 86% share of RES in the generation mix by 2050. On the end-use side, the share of electricity in final energy consumption would rise from the current value of 20% to almost 50% by 2050. The share of electricity consumed in industry and households would double. In transportation, it would rise from just 1% today to more than 40% by 2050 (IRENA 2019a).

Renewable electricity is currently the cheapest power option in most regions of the world. The global levelized cost of electricity (LCOE) from newly commissioned industrial photovoltaic installations decreased by 85% between 2010 and 2020, the cost of concentrated solar power (CSP) by 68%, onshore wind power by 56% and off-shore wind power by 48%. All commercially available solar and wind electricity are within (or even below) the cost of electricity from new fossil fuel-based power plants. It is also important to keep in mind that savings from larger scale, more competitive supply chains and further improvements in technology will further reduce the cost of renewable energy (IEA 2021a; IRENA 2020i).

For example, globally, the total cost of installing onshore wind farms will continue to decline over the next three decades, with an average cost in the range of $800–1,350 per kW by 2030 and $650–1,000 per kW by 2050, compared to a global weighted average of $1,497 per kW in 2018. For offshore wind farms, the average total installation cost will decrease in the coming decades to $1,700–3,200 per kW by 2030 and $1,400–2,800 per kW by 2050. The levelized cost of electricity (LCOE) for onshore wind power will continue to fall from an average of $0.06 per kWh in 2018 to $0.03–0.05 per kWh by 2030 and $0.02–0.03 per kWh by 2050. The LCOE of offshore wind power would decline from an average of $0.13 per kWh in 2018 to an average of $0.05–0.09 per kWh by 2030 and $0.03–0.07 per kWh by 2050. Technological innovations toward higher capacity turbines, taller masts and larger rotor diameters will raise efficiency in the same location. For onshore wind farms, global weighted average capacity factors would increase from 34% in 2018 to a range of 30%–55% in 2030 and 32%–58% in 2050. For offshore wind farms, the progress would be even greater, with capacity factors reaching a range of 36%–58% in 2030 and 43%–60% in 2050, compared to an average of 43% in 2018 (IRENA 2019).

As a result, RES is already the default option for capacity expansion in the power sector in almost all countries and dominates current investments. Within this group of energy sources, solar and wind technologies have consolidated their dominance, which, given the recent increases in fossil fuel prices, creates a promising economic outlook for renewable energy.

Wind energy has been used by humans for many years. There are reports of "Persian" windmills; moreover, windmills (with a vertical axis of rotation) provided irrigation for fields in China and Egypt. These are historical facts dating back to before the birth of Christ. In Europe, reliable records say windmills of the trestle type (with a horizontal axis of rotation and four blades) were encountered in Normandy in the 12th century. By the mid-19th century, windmills were very popular and widely used to power various types of equipment. It is estimated that about two hundred thousand windmills were in operation in Europe (Lubośny 2006). The design featuring a wind wheel consisting of multiple blades and a distinctive wind vane is an American standard found in the past in the United States almost everywhere outside urban areas. The beginning of the end of the windmill era was associated with the advent of steam engines. The end of this era came with increasing electrification, especially when it covered sparsely populated areas.

Already in the late 19th century, there were attempts to use wind energy to generate electricity. Thus, paradoxically, windmills found use in supporting the electric power system, which, as indicated above, had previously contributed to their demise. The beginning of the 20th century brought, among other things, the solutions of Danish engineer Poul La Cour, in the form of commercially available wind turbines driving DC generators with the output of 10–35 kW (Ackermann 2009). Interestingly, the same designer pioneered the production of industrial quantities of hydrogen by electrolysis. Rather than ecology, the motivation for undertaking the aforementioned work was, of course, the search for solutions to provide power to island systems, which were a natural step in the development of an extensive electric power system. As stated in Lubośny (2009), the transition from these solutions to modern turbines, the power of which already exceeds 10 MW, was neither easy nor quick, as it took more than 100 years and consisted of a series of spectacular successes and failures, both technical and technological.

Opponents of wind power claim that its development was the result of an obligation administratively imposed by European Union directives to achieve a certain level of energy production from RES. Although one can partially agree with such a statement, the development trend of wind turbines, which in 2007–2009 turned from linear to parabolic, resulted from the involvement of China, the United States and other non-European countries. These countries were not subject to the Union's guidelines, but saw wind power as the prospectively cheapest source of clean energy. This statement is now undisputed, although the photovoltaic boom has caused some investor groups to shift their interests toward solar power.

The history of wind power development in Poland is rife with dramatic twists and turns resulting from legal and political decisions. The emergence of a market for green certificates guaranteeing wind turbine energy prices at an attractive level resulted in a huge increase in interest in wind turbine investments and the submission of applications to distribution system operators (DSOs) and transmission system operators (TSOs) for connection conditions for 40–50 GW of capacity by 2010. Today, this trend can be assessed as artificially stimulated and inflated; however, it constitutes a natural measure of the potential of Polish onshore wind power. The development of wind power in Poland came to a drastic halt in 2016 as a result of the introduction of the so-called Distance Law (10H) (Distance Act 2016). It is also worth mentioning that the collapse of the green certificate market and the view that persisted for several years denied the possibility of safe operation of the power system with a share of wind turbines greater than 10 GW. The small increase in installed wind power capacity (in 2021 it exceeded 6.5 GW) is the result of the auction system and the implementation of favorable administrative decisions obtained by some facilities even before the 10H Law came into force. However, these opportunities have now been virtually exhausted. In recent years, there have been announcements of liberalization of the requirements of the Distance Law, which, as of today, appear to be real. Planners' forecasts indicate that in 2040, onshore wind energy in Poland will reach a capacity of 20 GW, so it has finally been classified as one of the basic technologies creating the country's energy mix.

Meanwhile, globally, 2020 turned out to be a record year for wind power. Despite the pandemic, the capacity increased by 93 GW, nearly a third more than in the record year of 2015. It is worth noting that 93% of the new installed capacity corresponds to onshore wind farms. Offshore, 6.1 GW of wind turbines were installed. The leaders in the construction of new farms are China (52 GW of new capacity) and the United States (16 GW). Among European countries, investment processes in wind power were much more modest in 2020: the Netherlands – 2 GW, Germany – 1.6 GW (Olszowiec 2021). The total installed capacity of wind turbines in the world exceeded 750 GW, again with China leading the position (290 GW). In Europe, Germany (53 GW), Spain (27 GW) and the UK (24 GW) have the largest installed capacity. In virtually all European countries, the share of wind power in the overall balance of electricity consumption is steadily increasing. According to data from the ENSTO-E organization, this indicator in 15 EU countries has exceeded 10%, and the average for the entire EU is 15% (Denmark 48%, Ireland 36%, Germany and the UK 27% each, Sweden 20%). For Poland, the indicator is 9%. Wind turbine manufacturers also experienced record achievements in the past year. The Haliade-X turbine (made by GE), which has been tested since late 2019, has reached 14 MW, and 12 MW turbines are ready for offshore applications.

It is important to bear in mind the projections that onshore and offshore wind power would generate more than a third (35%) of total electricity demand in 2050, becoming the main source of generation. This represents a more than threefold increase in the global cumulative installed capacity of onshore wind power by 2030 (to 1,787 GW) and ninefold by 2050 (to 5,044 GW) compared to the installed capacity in 2018 (542 GW). For offshore wind, global cumulative installed capacity would increase almost tenfold by 2030 (to 228 GW) and further by 2050, with total offshore installed capacity approaching 1,000 GW in 2050.

It is estimated that the wind industry could employ 3.74 million people by 2030 and more than 6 million people by 2050, nearly three times and five times the 1.16 million jobs in 2018, respectively. A friendly policy framework is needed to maximize the results of the energy transition.

The study presented is the authors' voice in the ongoing national discussion on the energy transition and the role and contribution of renewable energy sources. The presented work focuses its attention on wind energy installed onshore. In the aforementioned discussion, the role of the 10H law is prominent. The current form of this legislation has led to a significant decrease in the dynamics or outright blocking of the development of this branch of renewable energy sources, and calls into question the real possibility of Poland meeting its climate commitments to the international community.

It should be borne in mind that the implementation of the energy transition, especially its form and pace, will largely depend on a favorable policy framework, including legal solutions.

The content and suggestions cited in this monograph should be helpful in the implementation of the forthcoming amendment to the aforementioned law and other legislation relevant to onshore wind energy development. In the text of the monograph, the authors often refer to legal acts published in Poland and issued by the Polish authorities. However, the problems raised are universal and can also be considered in other countries, even if the regulations in force in them are slightly different.

1 Wind Turbines – Types of Impacts and Methods to Assess Them

1.1 ACOUSTIC IMPACTS

Noise is one of the most harmful environmental factors affecting human health. The nuisance and harmfulness of noise mainly depend on its frequency, duration of exposure, or intensity and content of inaudible components, as well as the personal characteristics of the recipient (Lis et al. 2015). Individual characteristics, such as noise sensitivity, privacy issues and social acceptance, benefits and attitudes, local situation and wind farm planning conditions, also play a role in reported nuisance. Environmental noise is a recognized problem that appears to be growing both due to increased exposure to noise from existing sources, such as traffic, and due to the introduction of new noise sources, such as wind turbines (Mohamed 2014). Sound levels can be measured, but as with other environmental problems, the public's perception of the acoustic impact of wind turbines is partly subjective.

1.1.1 DESCRIPTION OF THE PHENOMENON

The decades-long history of investment in wind power shows that it can be a low-cost source of renewable energy, while on the other hand, it has an impact on the environment. During the operation of a wind turbine, noise is emitted in the audible frequency range (the range of 20 Hz–20 kHz), in addition to infrasound noise, commonly referred to as inaudible (in the range of 0.1–20 Hz). Its propagation depends on many factors. Computer simulations, approximating (using specialized software) the influence of these factors on the parameters of propagating sound, and actual acoustic measurements are the basis of any documentation necessary to obtain a permit for the construction of a wind farm. The calculations are illustrated graphically on noise maps and determine the extent of the noise, which makes it possible to assess the noise in the nearest acoustically protected areas. Both the environmental impact assessment report and the decision on the environmental conditions of such a project depend on a well-prepared acoustic report, from which the degree of annoyance of the wind farm's operation to local residents should be determined. One of the main elements that can affect such nuisance is the distance between the investment and the human habitat. In this study, much attention will be paid to determining the so-called "minimum acceptable distance," after which the probability of the impact of a wind farm on human health may already be significant.

DOI: 10.1201/9781003456711-2

1.1.2 BASIC TERMS

Noise – unwanted sound that can be a nuisance or harmful to health. Through the air, noise in the environment affects the human auditory and other organs. Noise is a collection of sounds with different frequencies and different sound pressure values.

Acoustic vibration – oscillatory motion of the particles of an elastic medium relative to the equilibrium position. This disturbance involves the transfer of mechanical energy by vibrating particles of the medium (compression and expansion) without changing their average position. The elastic medium can be a gas, liquid or solid.

Acoustic wave – the propagation of acoustic vibrations that cause an auditory sensation (sound wave) and vibrations with frequencies beyond the human auditory range.

Division of sounds by frequency:

- infrasound, range 0.1–20 Hz,
- audible sounds, range 20–20,000 Hz,
- ultrasound, range above 20,000 Hz.

In practice, the frequency ranges shown are expanded or reduced depending on the applicable noise assessment methodology. In recent years, another group of sounds – low-frequency noise (LFN) – has been identified. As yet, there is no uniform international standard defining the range of LFN. Many countries adopt their own LFN ranges, in the frequency range of 10–200 Hz. In Poland, there are no regulations governing permissible levels of low-frequency noise (LFN).

Due to the variation in sound intensity over time, the following types of noise may be distinguished:

- steady-state noise – the sound level varies no more than 5 dB during observation,
- fluctuating noise – the sound level during observation varies more than 5 dB,
- impulsive noise – consists of one or a series of sound impulses, each lasting less than 1 second.

Sound power – a measure of the amount of energy emitted by a source per unit time. It is the basic quantity, determining the emission of the source.

Sound intensity – the value of sound power, per unit area, perpendicular to the direction of propagation of the acoustic wave.

Decibel (dB) – a logarithmic unit used when comparing linear quantities over a very wide range of values. In acoustics, the pressure changes of 0.00002 (hearing threshold) to 100 (pain limit) Pa are compared, which gives a range of 0–134 dB. Converting the linear unit (Pa) to logarithmic (dB) follows the formula:

$$L_P = 10\log\frac{p^2}{p_0^2} \qquad (1.1)$$

where

L_P – sound pressure level, dB,

p – average measured sound pressure, Pa,

p_0 – reference pressure ($2 \times 10^{-5} = 0.00002$), Pa; threshold pressure, corresponding to the sound pressure for a tone with a frequency of 1,000 Hz, at which the hearing sensation begins.

Equivalent sound pressure level ($L_{eq,T}$) – the value of a continuous, fixed sound that has the same mean square sound pressure level over a specified time interval (T) as the sound under consideration, which varies over time:

$$L_{eq,T} = 10\log_{10}\left(\frac{1}{T}\int_0^T \frac{p^2}{p_0^2}\,dt\right) \tag{1.2}$$

The permissible levels of noise are determined in the environment using the equivalent sound level (Regulation – noise 2007), which is an indicator for assessing the impact of noise on the environment.

Sound power level, in dB – an acoustic measure of the amount of acoustic energy that is radiated from a sound source and shown as a ratio to the reference sound power: $W_0 = 10^{-12}$ W

$$L_W = 10\log_{10}\left(\frac{W}{W_0}\right) \tag{1.3}$$

Spectral analysis – illustrates the distribution of simple sounds that make up the level of a complex sound. The basic method of spectral analysis is the Fourier transform. However, in environmental noise impact assessments, the spectral distribution is determined by filter methods. One-third and octave filters, with a fixed relative bandwidth, are used most commonly.

Sound can be divided into frequency bands to look more closely at the distribution of sound pressure across the frequency spectrum. Most often, the audible frequency spectrum is divided into 10 bands that are one octave wide. This means that the upper and lower band limits remain in a ratio of 2:1. The center frequencies of the octave bands are given in Hertz. The information provided by octave band analysis is often not detailed enough. A more detailed breakdown can be obtained by octave band analysis, which divides the audio signal into bands which are one-third octave wide. Then, one octave consists of three one-third octaves, and the audible frequency spectrum is divided into 30 one-third octave bands. Table 1.1 shows the center frequencies of the octave and one-third octave bands, as well as their widths.

Weighting curves – approximate the value of energy carried by sound to the sensitivity of the human ear. They are used in noise exposure assessments. The most commonly used are weighting curves A for moderately loud audible sounds, G for infrasound, as well as C for assessing the levels of loud audible sounds and peak sounds at workplaces. The A weighting curve is almost an inverted curve, depicting the hearing capabilities of the human ear, depending on the frequency of sound (MHT). Humans hear best in the frequency range of 500–10,000 Hz, and less well outside this range. Correction is taken into account by corrections subtracted or added to the measured values in the one-third octave bands. The correction values are illustrated in Figure 1.1.

It should be noted that the thresholds for auditory perception of infrasound are at high levels, which means that humans perceive low sound frequencies very poorly. The thresholds for frequencies up to 100 Hz are summarized (Watanabe 1990) in Table 1.2.

TABLE 1.1
Center Frequencies and Bandwidths of Octaves and One-Third Octaves

Octave bandwidth	31.5			63			125			250			500		
Lower limit	22	28	35	44	57	71	88	113	141	176	225	283	353	440	565
One-third octave bandwidth	25	31.5	40	50	63	80	100	125	160	200	250	315	400	500	630
Upper limit	28	35	44	57	71	88	113	141	176	225	283	353	440	565	707

Octave bandwidth	1,000			2,000			4,000			8,000			16,000		
Lower limit	707	880	1,130	1,414	1,760	2,250	2,825	3,530	4,400	5,650	7,070	8,800	11,300	14,140	17,600
One-third octave bandwidth	800	1,000	1,250	1,600	2,000	2,500	3,150	4,000	5,000	6,300	8,000	10,000	12,500	16,000	20,000
Upper limit	880	1,130	1,414	1,760	2,250	2,825	3,530	4,400	5,650	7,070	8,800	11,300	14,140	17,600	22,500

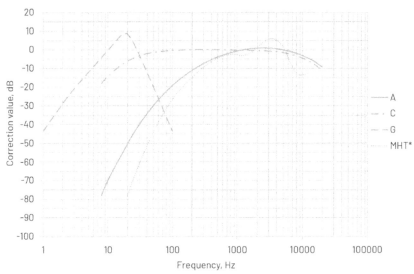

* MHT - inverted mean hearing threshold curve

FIGURE 1.1 Weighting curves A, C and G and inverted mean hearing threshold curve.

TABLE 1.2

Human Hearing Thresholds

Frequency (Hz)	4	8	10	12.5	16	20	25	31.5	40	50	63	80	100
Level (dB)	107	100	97	92	88	79	69	60	51	44	38	32	27

Noise emissions from a wind turbine – a wind turbine emits aerodynamic noise, created by air flow, and mechanical noise, created by the interaction of solids, usually machine parts. Aerodynamic noise is generated mainly during the flow of air down the rotor blades and during the passage of the blade near the turbine support tower. Mechanical noise is generated by the equipment installed in the nacelle on the tower. The acoustic power of wind turbines, given as the sum of aerodynamic and mechanical noise, always depends on the wind speed due to the nature of the device's operation. Exemplary acoustic power values of a wind turbine, depending on wind speed, are shown in Table 1.3 and Figure 1.2. The table shows wind speeds at two heights above ground level (10 and 98 m). Wind speeds and corresponding sound power levels are given in different ways by turbine manufacturers. Most often, in the technical documentation, the relationship is presented as a graph or a table. Many manufacturers also provide the data on sound power levels broken down into values in octaves and one-third octaves, which enables to carry out calculations of the noise range taking into account the influence of the ground using the general method, in accordance with the current PN-ISO 9613-2 standard (PN-ISO-9613 2002). In some technical documentation of turbines, the maximum sound power level is given for two wind speeds, i.e., at the level of 10 m and the corresponding wind speed at the level of the rotor axis of the wind turbine.

TABLE 1.3

Sound Power Levels of an Example Wind Turbine as a Function of Wind Speed

		3	4	5	6	7	8	9	10 to max
Wind speed at a height of 10 m, m/s		3	4	5	6	7	8	9	10 to max
Wind speed at the height of the rotor axis (98 m), m/s		4.3	5.7	7.2	8.6	10.0	11.4	12.9	14 to max
Spectral analysis of octave, frequency, Hz	31.5	68.7	68.5	72.9	76.9	79.3	80.2	80.5	80.8
	63	77.8	77.6	82.2	86.4	88.9	90.0	90.1	90.3
	125	82.1	82.4	87.3	91.6	94.7	95.3	95.2	95.3
	250	85.6	86.1	91.3	96.0	99.9	99.0	98.1	97.6
	500	88.4	87.9	93.4	98.6	102.2	101.6	101.3	100.5
	1,000	87.9	87.5	92.1	96.6	99.0	100.1	101.0	101.8
	2,000	84.4	85.9	90.7	94.7	96.9	97.7	98.1	98.5
	4,000	74.5	77.7	93.8	88.6	91.5	91.7	90.7	89.8
	8,000	55.1	56.9	63.4	69.6	73.6	73.0	70.8	70.5
	16,000	8.4	12.7	18.8	24.2	28.6	27.5	28.4	27.7
The total sound power level L_{WA}, dB		93.4	93.6	98.6	103.3	106.5	106.5	106.5	106.5

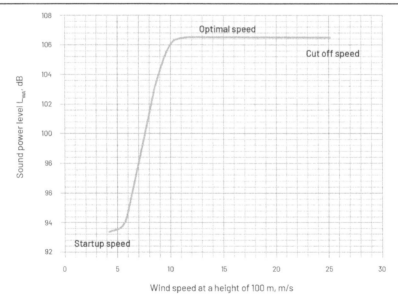

FIGURE 1.2 Dependence of sound power on wind speed.

Characteristic acoustic phenomena associated with the operation of wind turbines are as follows:

- The sound power level of the types of turbines available on the market is diversified and variable during operation of the device, depending on the wind speed. It increases along with wind speed and is virtually constant after reaching a certain optimum speed.

- Wind speed increases with the change in height above ground level, i.e., the higher the altitude, the greater the wind speed. This fact is important for the productivity of wind turbines and the possibility of conducting control noise measurements around the turbines.

In order to conduct an assessment for the most acoustically unfavorable case of wind turbine operation possible, measurements would have to be made at the wind speed corresponding to the maximum sound power levels of the turbine and the minimum wind speed, at the height of the measurement point. Such a situation corresponds to the wind profile marked with the letter M in Figure 1.3. The figure was prepared on the basis of literature data (Kariniotakis 2017; Van den Berg 2008) and own observations.

- The background noise level in the environment changes with the wind speed, increasing with it (Figure 1.4).

The study results (Acoustica 1997) show that the contribution of background noise, in the measured total noise level, tends to increase with the wind speed. At higher wind

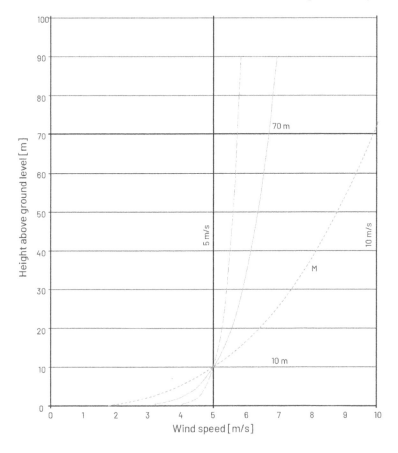

FIGURE 1.3   Examples of vertical wind profiles.

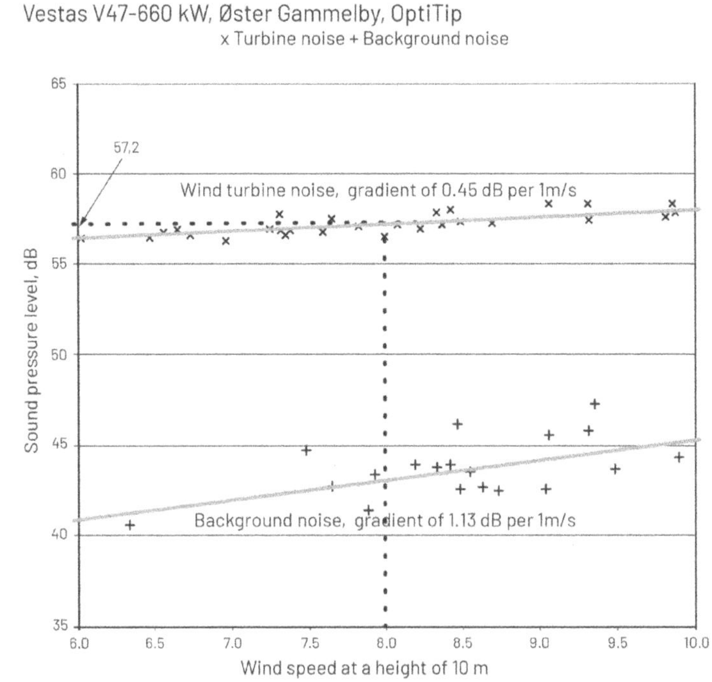

Vestas V47-660 kW, Øster Gammelby, OptiTip
x Turbine noise + Background noise

FIGURE 1.4   Variability of background noise with change in wind speed.

speeds, close to reaching the maximum sound power level of the turbines, the background level at the measurement point becomes comparable to the noise level from turbine operation. It is then found that the noise from the operation of the turbines is indistinguishable from the background noise, which makes it impossible to assess the impact of the wind farm by measurement. The absence of exceedances of permissible noise levels is then found on the basis of simulations (by calculation). One of the important reasons for this situation is the considerable distance of the source from the control measurement points (more than 400 m), located near the nearest residential buildings. An additional factor contributing to the increase in background noise levels in the vicinity of the measurement points is the greater "roughness" of the terrain around the development. The reason for this is the presence of trees, shrubs, buildings and other objects in the vicinity of the measurement point. This phenomenon has been confirmed in the paper by Bullmore (2009) and in dozens of acoustic measurement assessments, carried out by the authors of this study, for large wind farms in Poland.

1.1.3   EMISSION SITES AND MEANS OF ACOUSTIC WAVE GENERATION

Wind turbines generate both mechanical and aerodynamic sound. However, current technology makes mechanical noise less important, as under normal operating conditions it decreases below the level of aerodynamic noise (Pedersen et al. 2006). Developments in technology have made wind turbines much quieter, but the sound they emit is still an important placement criterion. Sound emissions from wind turbines are one of the better studied areas of environmental impact in wind power (Rogers et al. 2006).

The widespread use of classic wind turbines, i.e., with a horizontal axis of rotation, makes their acoustic parameters readily available. The manufacturer of turbines is obliged to provide the acoustic characteristics of their product, for example, as part of CE mark certification, where a measurement procedure is used for acoustic evaluation that includes the determination of, i.e., the sound power level, according to IEC 61400 (IEC 61400-11 2012). The number of wind parks already installed means that wind turbine noise has become a field of scientific research, the results of which are discussed at many international conferences (e.g., Wind Turbine Noise, since 2005). One of the first studies that comprehensively presented information on the generated noise of wind turbines is the work of Wegner et al. (1996), which presents the sources of noise emissions of turbines and the components of the acoustic field, which are relevant in the classical view of SOURCE–PROPAGATION ROUTE–RECIPIENT and the sound power levels of individual turbine components (Figures 1.5 and 1.6).

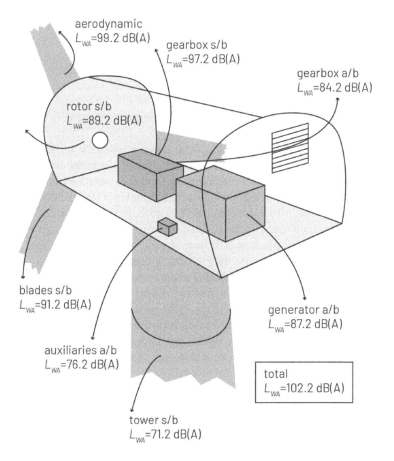

FIGURE 1.5 Sound power level of the components of a classical wind turbine with electric power output 2 MW (Wegner et al. 1996).

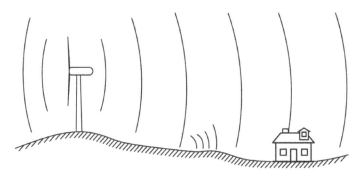

Noise generation   Propagation   Reception

- Aerodynamic sources   - Distance   - Ambient noise
- Mechanical sources   - Wind gradient   - Being indoors/outdoors
                          - Absorption   - Building vibrations
                          - Terrain

FIGURE 1.6 Components of the sound field in terms of the principle: Noise source–Propagation path–Recipient (Wegner et al. 1996).

Wind turbine noise has been divided in terms of acoustic emission sources into the following (Wegner et al. 1996):

a. mechanical noise – originating from the nacelle (*generator, gearbox, nacelle*)
b. aerodynamic noise – from the movement of rotating blades (disturbance of the elastic medium at blade tips, Karman vortices, air cavitation, changes in the pressure of the elastic medium during the passage of the blade past the tower) (Figure 1.7)

Aerodynamic noise results from the operation of the blades, which cause acoustic emissions both in the infrasound range (inaudible for the human ear) and in the audible range, the characteristic sound of the passage of the blade through the tower, i.e., swoosh.

Mechanical noise represents the dominant components of the amplitude-frequency spectrum in the audible range, which are characterized by components in the range of more than 300–500 Hz, due to sufficient attenuation of the sound wave by air, which is not a significant impact in acoustically protected areas. Most wind turbines have soundproofing of the nacelle itself, which further reduces the noise from mechanical components of wind turbines (Figure 1.8).

1.1.4 Acoustic Wave Propagation in Space

Noise emission forecasting at the design stage (e.g., environmental report, acoustic optimization of the farm) is performed based on the calculation method recommended for industrial noise in Directive 2002/49/EC of the European Parliament and of the

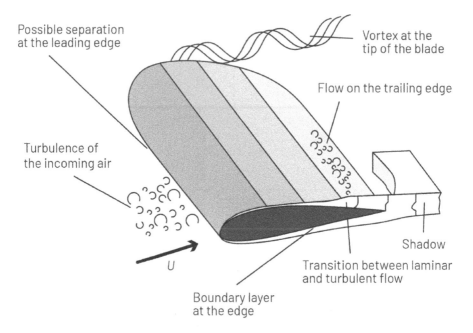

FIGURE 1.7 Source of aerodynamic noise generation during the operation of standard wind turbines (Wegner et al. 1996).

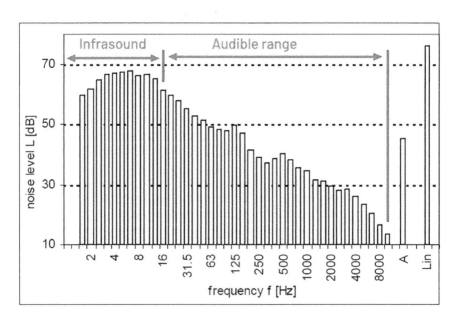

FIGURE 1.8 Sound spectrum of the Vestas V-80 Turbine (Golec et al. 2005), with changes implemented by the author.

Council of June 25, 2002 (Directive 2002/49 EC). This method is described in the Polish standard PN-ISO 9613-2:2002 *Akustyka, Zmniejszanie propagacji dźwięku na otwartej przestrzeni, Ogólna metoda obliczeń* (PN-ISO-9613 2002) (Acoustics, Reduction of sound propagation in open space, General method of calculation).

For the considered computational method, the noise ranges from proposed wind turbines are determined through computational models including:

- digital model of the project site,
- location of planned wind turbines,
- location of areas under acoustic protection.

The digital terrain model is created based on design maps, which should be consistent with the master and topographic maps. Wind turbines, as sources of noise emissions, are included in the computational model as so-called surrogate point sources located in the simulation model at the height of the nacelle (Makarewicz 2011).

According to the cited standard, the propagation of a sound wave in the environment is characterized by attenuation (A) of the acoustic energy radiated from the noise source to the external environment. This attenuation is the resultant magnitude of several components relating to various physical phenomena accompanying sound propagation and is expressed by the following formula:

$$A = A_{div} + A_{atm} + A_{gr} + A_{bar} + A_{misc} \tag{1.4}$$

where

A_{div} – attenuation resulting from the so-called "geometric divergence," i.e., spherical propagation of an acoustic wave from a point source of sound;

A_{atm} – attenuation due to atmospheric absorption;

A_{gr} – attenuation resulting from interaction with the surface over which the sound propagates;

A_{bar} – attenuation due to the presence of obstacles in the path of sound propagation between the source and the observation point;

A_{misc} – attenuation due to other phenomena accompanying sound propagation, including absorption during propagation through areas of high vegetation, areas of dense buildings or industrial areas.

Assessments of the acoustic impact of wind farms, carried out at the stage of designing their location, are based on acoustic modeling. Practically all over the world, the methodology presented in ISO 9613-2 (PN-ISO-9613 2002) is used for calculations, whereas the Nord2000 (DELTA 2002) methodology is approved in Scandinavian countries. The latter can be considered as an extension of the ISO 9613-2 methodology, allowing for much greater modification of acoustic wave propagation conditions and analysis of cases other than just the worst case, which is calculated by using the ISO methodology. When evaluating noise with the ISO (PN-ISO-9613 2002) methodology, it is important to define the relevant calculation parameters (Table 1.4), which may vary from one country to another due to the adopted specific guidelines in force in a particular area or country.

TABLE 1.4

The Most Relevant Parameters for Calculating Wind Farm Noise Range

Calculation Parameter	Typical Settings
Sound power level	Maximum or set for a defined wind speed (usually 6, 8 and 10 m/s measured 10 m above sea level). Nowadays, it is most commonly entered as a spectral power distribution. In the past, single-number values were used.
Ground influence G	Defined for individual types of terrain: Water – $G = 0$, Land – $G = 0.4$–0.7. or Fixed attenuation value: Land – 3 dB, Sea – 1.5 dB.
Tonality	Application of an additional +5 dB correction if tonality is present.
Effect of meteorological conditions on propagation (C0)	Not applied in practice.
Altitude at which evaluations are conducted	Typically 4 m above sea level. Values of 5 and 1.5 m above sea level are also encountered.
Assessment indicators	Usually L_{AeqT}, L_{DEN} and L_{90} are used as well.

The use of such a modeling approach to determine noise ranges enables to obtain reliable sound levels caused by the operation of wind turbines, which will be highly comparable to those obtained from noise measurements as part of post-implementation studies or acoustic monitoring. Of course, this compatibility will be possible due to the features of turbine operation noise modeling described in Section 1.1.5.

1.1.5   RESEARCH, DESIGN AND ACOUSTIC OPTIMIZATION OF WIND TURBINES

The noise impact is an important component of the environmental impact of wind farm operation. Noise ranges from operating turbines depend mainly on the speed of the turbines, which is the effect of instantaneous wind speeds at the height of the nacelle. The magnitude of noise at residential buildings depends on the direction of the wind (whether the noise is carried "downwind" or "upwind") and the seasons, which is determined by the conditions of absorption by the soil and the size of the acoustic background – for the winter months, the sounds of animate nature and the noise of leaves are absent, for the other months it is a significant component of environmental noise.

Thus, accepting the thesis that noise ranges from operating turbines depend on a number of factors, it is important to determine the sound levels at acoustically protected areas for these different variants of wind farm operation.

The essential and decisive parameters affecting the amount of noise from operating turbines relative to residential buildings where people live include:

a. the function of the acoustically protected area and the limit values corresponding to the given function (the lowest are 40 dB at night and 50 dB during the day for single-family residential development),

b. the type of planned turbine, that is, its acoustic characteristics:
  - its maximum sound power level and power level for a given wind class,
  - the height of the turbine, as the height of the nacelle and the length of the blades,
  - the type of blades in acoustic terms, such as those with serrated trailing edges to reduce their aerodynamic noise,
c. the number of turbines around a particular residential building (for a single turbine, the distance from the development will be much smaller than for a larger number of turbines at similar distances from the residential development),
d. the characteristics of the area between the turbines and the residential buildings, which has an impact on the determination of the coefficient of absorption of sound waves by the soil and the determination of the roughness of this ground at different times of the year,
e. the directionality of the wind turbine, i.e., the difference in sound levels between positioning the turbine sideways and forward or backward relative to the residential building.

The sum of all the abovementioned factors gives a reliable estimate of the amount of noise, which directly translates into the determination of a safe distance between the wind turbine and acoustically protected residential buildings.

Performing acoustic calculations taking into account the criteria shown above makes it clear that the distance for which the permissible noise values at residential buildings are met is within a very wide range. Therefore, the main conclusion to be drawn from the stage of designing wind farms is that a reliable determination of the safe distance between a turbine and a residential building should be made every time on the basis of noise calculations, i.e., calculations for given assumptions (number and type of turbine, permissible values, type of terrain – its roughness, etc.). It is also clear that all acoustic analyses should be performed for the variants least favorable from the point of view of noise ranges.

The design stage involving the determination of the safe distance between the turbine and the residential building should take into account the accuracy of the noise calculations at the design stage, the reliability of the calculations, which at a later stage can be compared with the results of physical noise measurements, and the uncertainty, i.e., the risks that may occur and cause the obtained results of noise measurements that will be made after the wind farm is put into operation to be higher than those shown at the design stage.

The accuracy of noise calculations at the design stage and the scatter of results from field noise measurements are a serious issue when designing a wind farm in terms of acoustics. The resolution of acoustic programs and the accuracy of the results obtained is 0.1 dB. This means that if the established noise limit for residential development is 40 dB, and the sound levels obtained from the simulation program's calculations at the reference point, e.g., at the border of the protected area, are 40.1 dB, noise is formally exceeded. The solution to such a situation at the design stage is to reduce the sound power level by using the turbine's silenced mode. At the design stage, the reduction of the sound power level is achieved by subtracting, for example, 1 dB or more from the maximum sound power level of the turbine – using so-called "mods" of turbine operation. In technical terms, such turbine noise reduction (subtracting a specific number of dB) involves physically reducing the turbine's speed, which further results in lower turbine operating noise. Turbine noise reduction ranges of so-called "mods" have effectiveness in the range of 1 dB to as much as 3 dB for a given "mod." This means that by reducing a turbine noise by one mod, its sound power level can be reduced by up to 3 dB. In addition, each turbine has a greater number of noise reduction modes ("mods"), which means that a turbine operating at its maximum sound power level can be suppressed by up to 5–7 dB. The number of noise reduction modes, their range in dB, is different for different turbine manufacturers and depends on the features of a given turbine design, i.e., stiffness, resistance to generator vibration, type of blades, etc.

However, measurement practice shows that the variability of sound levels during physical noise tests of wind turbines can reach 5 dB and more. This is mainly the result of local wind gusts and the influence of the acoustic background, such as household sounds, sounds of animate nature, or other noise sources occurring in the area of the residential development under study, i.e., traffic. It is important to know what sounds constitute noise from the operation of the turbines and what is the acoustic background during noise measurements.

In the best case, if an exceedance of 0.1 dB (a value of 40.1 dB) is obtained from noise calculations, then 1 dB noise reduction has to be applied, which results in a reserve of 0.9 dB for residential development. In the worst case, 3 dB noise reduction has to be applied for an exceedance of 0.1 dB, thus creating a significant "noise reserve." Such measures automatically make the reliability of noise analyses high, and the noise levels at acoustically protected areas are always overestimated (inflated), relative to the actual noise levels, which are physically measured at daytime and nighttime. Elements that increase the reliability of the results or, in other words, their uncertainty, i.e., the estimated difference between the levels obtained from noise calculations and from post-implementation measurements, are two additional factors:

a. Equal operation of all turbines – noise analyses assume that all turbines operate stably and without losses due to the influence of other turbines. In practice, however, over the entire farm area, if even the wind strength is such that the turbines operate at maximum speed, all turbines never operate at 100% capacity. This is due to the so-called Betz's law, which shows that the wind speed behind the turbine compared to that in front of the turbine is about 30% lower. This 30% of the wind force is converted into rotational motion of the blades. These losses, despite the considerable distances and alignment of the turbines with respect to each other, result in the fact that not all turbines operate at maximum speed – with maximum sound power levels.

b. Impact of acoustic background (mainly wind noise) and other sources of noise emissions. The effect of acoustic background on noise levels from turbines is an element that is often overlooked in noise calculations (at the farm design stage). Generally speaking, the acoustic background is all the sounds in the vicinity of the measurement point that are not part of the noise source under study, that is:

- turbine noise and wind background at similar levels;
- turbine noise and the road;
- turbine noise and another industrial source;
- cumulative effect with another farm.

In a situation where the acoustic background around the residential building under study is also shaped by other sources than just noise from the wind turbine, the size of the correction taking into account the influence of the background can be as high as 2.5 dB. This follows directly from the relationship for calculating the level of noise emissions in the environment taking into account the correction for the acoustic background:

$$L_{Aek} = 10\log\left(10^{0,1L_{Asr}} - 10^{0,1L_{At}}\right) \tag{1.5}$$

where
L_{Asr} – the average measured sound level A in [dB]
L_{At} – the average background sound level in [dB]

In other words, it is important to know that the noise calculations presented at the design stage are usually inflated, relative to those that are later physically measured. If the acoustic design of a wind farm has been reliably carried out then it gives confidence that the measured noise levels during the operation of the wind farm will not exceed acceptable levels and will not pose a threat to the people living there.

1.1.6 Noise Assessment Indicators, Threshold Values

In Europe and around the world, assessments of the environmental impact of wind farm noise are carried out by various methods (Raman et al. 2016; Fredianelli et al. 2019; Davy et al. 2020). Determination of permissible levels is carried out mainly by using three methods. The first is carried out by establishing absolute indices, usually dependent on the time of day and land use. The second is by establishing relative indicators, which depend on the current state of acoustic conditions in the analyzed area. The third is a combination of the previous two, that is, using relative and absolute indicators together. Sometimes a 5 dB correction for the tonal nature of the source is also applied (Hansen et al. 2017). The WHO recommends a permissible long-term noise level L_{DWN} for wind turbines equal to 45 dB (World Health Organization 2018). In Poland, the absolute indicators specified in the regulations of the Minister of the Environment (Journal of Laws 2014, item 112) (Regulation – noise 2007) are used. Wind turbine noise assessments are based on analyses of the sound signal corrected by the A correction curve (Figure 1.1). Daily equivalent levels commonly used for assessment are, e.g., $L_{Aeq,D}$/daytime and $L_{Aeq,N}$/nighttime, e.g., in Poland as well as annual indices based on them, L_{DWN}, L_N e.g. in the Netherlands or L_R (similar to L_{Aeq}) e.g. in Denmark. Countries such as the UK (ETSU-R-97 1996), New Zealand or Australia, where statistically based indicators are used, e.g., LA90, LA50, constitute exceptions. A few countries, such as France, use relative evaluation criteria checking the noise allowance above the acoustic background, which, due to the variable acoustic background, during the operation of wind turbines, can lead to ambiguous assessments. In addition, in France, limits can still be increased by 1, 2 or 3 dB when the exceedance only occurs for 4–8, 2–4 hours or 20–120 minutes, respectively. The values of indicators compiled in Table 1.5 were developed on the basis of literature (Hansen et al. 2017; Davy et al. 2018).

In each country, the assessment indicators are closely linked to the employed measurement methodology. The definitions of a protected area are also different. For example, in rural Poland, there are homestead, single-family and multi-family buildings, which have different values of L_{Aeq} indicators. Thus, it is not possible to simply compare with each other the values of indicators used in different countries. For example, in Sweden the L_{Aeq} indicator is determined for a specific wind speed, while in Poland a range of acceptable wind speeds during measurements is given.

TABLE 1.5

Selected Permissible Noise Levels and Audible Noise Assessment Indicators Used Worldwide

Country or Region	Assessment Indicator	Rural Areas		Residential Areas	
		Daytime	Nighttime	Daytime	Nighttime
Australia (Victoria, South Australia)	$L_{A90,\,10\,min}$	40	40	40	40
Australia (Queensland)	L_{Aeq}	37	35	37	35
Australia (New South Wales, Western Australia)	L_{Aeq}	35	35	35	35
Belgium (Flanders)	L_{Aeq}	48	43	44	39
Belgium (Wallonia)	L_{Aeq}	45	45	45	45
Canada (Alberta)	L_{Aeq}	–	40	–	43–46
Canada (British Columbia, Manitoba, New Brunswick)	L_{Aeq}	40	40	40	40
Canada (Ontario)	L_{Aeq}	40	40	45	45
Denmark	L_{Aeq}	42–44	42–44	37–39	37–39
France	L_{Aeq}	35	35	35	35
Germany	L_{Aeq}	50	35	50	35
Ireland	L_{Aeq}	40	40	40	40
Italy	L_{Aeq}	50	40	55	45
New Zealand	$L_{A90,\,10\,min}$	40	35	40	40
Norway	L_{DWN}	45	45	45	45
Sweden	L_{Aeq} for 8 m/s	35	35	40	40
Switzerland	L_{Aeq}	50	40	50	40
Netherlands	L_{DWN}	47	47	47	47
Netherlands	L_{N}	41	41	41	41
United Kingdom	$L_{A90,\,10\,min}$	35–40	43	35–40	43
USA (Georgia)	L_{Aeq}	55	55	55	55
USA (Minnesota, Wyoming)	L_{Aeq}	50	50	50	50
USA (Wisconsin)	L_{Aeq}	50	45	50	45
Poland	L_{Aeq}	50–55	40–45	50–55	40–45

Due to concerns expressed about the possible negative effects of low-frequency noise from wind turbines, a compilation of existing or proposed low-frequency noise limits from around the world has been made (Davy et al. 2020). Table 1.6 shows selected values of reference curves that have been used or are suggested for use in controlling general low-frequency noise in Europe. With the exception of the Polish and Danish nighttime values, which are based on A-curve corrected values, the values presented are close to the human ear's threshold values at 50 Hz and below. This means that the creators of these limits recognized that low-frequency noise close to the threshold of hearing can be annoying. Researchers' opinions are divided on the subject, and studies are still underway to rule out or confirm the potential health risks of low-frequency noise. In Poland, low-frequency noise, like infrasound, is not studied in any environment.

1.1.7 Tonality and Amplitude Modulation

A tonal noise (tonality) can be defined as one in which tonal frequency components are identified based on spectral analysis. The simplest way to determine the tonality of noise is to conduct sound pressure measurements in tertian bands. The presence of tones is determined by comparing sound pressure values in adjacent bands. Many regulations in various countries include a penalty for tonal noise up to 5 dB. It is generally accepted that any noise that contains distinguishable tones is more annoying than the noise of the same level without tones. In the case of wind turbines, it is necessary to establish an appropriate measurement methodology for tonal noise. The most

TABLE 1.6

Reference Curve Values Proposed Worldwide for Low-Frequency Noise Assessment

Frequency	Poland	Denmark	Germany	Netherlands	Sweden	United Kingdom	ISO 22
Hz	dB (A)				dB		
8	–	–	103	–	–	–	–
10	80.4	90.4	95	–	–	92	–
12.5	83.4	93.4	87	–	–	87	–
16	66.7	76.7	79	–	–	87	–
20	60.5	70.5	71	74	–	74	74.3
25	54.7	64.7	63	64	–	64	65
31.5	49.3	59.4	55.5	55	56	56	56.3
40	44.6	54.6	48	46	49	49	48.4
50	40.2	50.2	40.5	39	43	43	41.7
63	36.2	46.2	33.5	33	41.5	42	35.5
80	32.5	42.5	28	27	40	40	29.8
100	29.1	39.1	23.5	22	38	38	25.1
125	26.1	36.1	–	–	36	36	20.7
160	23.4	33.4	–	–	34	34	16.8
200	20.9	–	–	–	32	–	13.8
250	18.6	–	–	–	–	–	11.2

important issue is determining where to make measurements. Measurements can be conducted near wind turbines or at the receiver in an acoustically protected area (IEC 61400-11 2012). A measurement made at a wind turbine is not very conclusive, since tonality close to the source does not at all imply the appearance of tonal noise at a receiver generally 500 or more meters away, particularly since the level of background noise in a protected area can effectively mask tonal elements. On the other hand, it is clear that demonstrating the lack of tonality of a source will exclude the emergence of tonal components at a significantly distant acoustically protected area from that source. It seems ideal to conduct measurements near the source and look for tonal components at the receiver at the source's emission frequencies. However, to date, the methods for making and evaluating such a measurement have not been established.

Amplitude modulation of wind turbine noise is the periodic variation of sound emissions at the speed of the rotor blades passing the turbine's support tower at a frequency of 0.5–2.0 Hz. This means that the amplitude changes from a minimum to a maximum and again to the minimum as the rotor blades move away and then closer to the support tower. It has been observed that with increasing distance (more than 500 m) from wind turbines, this phenomenon turns into a fluctuation of the sound level (Zagubień 2018; Bonsma et al. 2017). Under certain meteorological conditions and changes in the blade attack account, several researchers have observed enhanced amplitude modulation. Some researchers consider the phenomenon of amplitude modulation as a cause of nuisance for the people living near wind turbines (Bonsma et al. 2017; Hansen et al. 2019). Although many researchers have made a lot of effort explaining the formation and impact of sound amplitude modulation, the cause of the phenomenon and the ways to possibly minimize its formation have not been clearly established (Oerlemans 2011; Sedaghatizadeh et al. 2017; Ecotière et al. 2019). It has been discovered that the variability of sound levels is lesser in the case of turbines with lattice support towers compared to tubular towers (Zagubień & Wolniewicz 2019) for upwind rotors. It is hoped that an understanding of the mechanism of sound amplitude variation will result in the development of sensor-controlled systems that appropriately change the blade's angle of attack to minimize acoustic effects.

1.1.8 ASSESSMENT AND MEASUREMENT OF NOISE EMITTED BY WIND TURBINES

In Poland, reference methods, applicable to industrial noise measurements, are used when measuring the noise coming from wind farms (Regulation – emissions 2021). The obtained measurement results are compared with the permissible noise levels specified in the Regulation of the Minister of Environment (Regulation – noise 2007). Both regulations (Regulation – noise 2007, Regulation – emissions 2021) provide the legal basis for assessing the impact of noise around wind farms. This methodology was used for measurements and simulations, the results of which are presented in this book.

In order to correctly design the distance of residential development from the planned wind turbine, it is necessary to adopt objective criteria that will allow determining whether the noise from the operating device is too high (exceeding the limits), or whether it is within the established norms and the operation of the turbine will not be a nuisance. The permissible sound levels in the external environment, determined by the Regulation of the Minister of the Environment (Regulation – noise 2007) are used for this purpose. According to this regulation, the permissible values of the A – weighted equivalent sound level, L_{AeqT}, for noise from objects and groups of sources other than roads and railroads are determined in time intervals equal, respectively, to eight, least favorable, daytime hours between 06:00 and 22:00, and a single, least favorable, nighttime hour, between 22:00 and 06:00.

The aforementioned regulation also defines categories of areas requiring acoustic protection (Table 1.7).

In practice, this means that noise emission exceedances, i.e., from wind turbines, occur when sound levels are higher than 55 dB during the daytime, for 8 consecutive hours of the turbine's operation (this applies to homestead, multi-family or commercial developments) and 50 dB for single-family developments. These 8 hours of daytime must be consecutive, i.e., it could be the time interval from 06:00 to 14:00 or 07:00 to 15:00, etc. In practice, this means that even if there is one louder hour in the daytime

TABLE 1.7
Permissible Noise Levels in the Environment

		Other Facilities and Activities Being Noise Source	
No.	Type of Area	L_{AeqD} (dB) Interval Equal to 8 Consecutive Least Favorable Hours of the Day	L_{AeqN} (dB) Interval Equal to 1 Least Favorable Hour of the Night
1	a. Protective zone "A" of a health resort b. Hospital areas outside the city	45	40
2	a. Areas of single-family housing development b. Areas of buildings connected with permanent or temporary residence of children and young people[a] c. Areas of nursing homes d. Hospital areas in cities	50	40
3	a. Areas of multi-family housing and collective residence development b. Areas of homestead development c. Recreational and leisure areas[a] d. Residential and service areas	55	45
4	Areas in the downtown zone of cities with a population of more than 100,000[b]	55	45

[a] If these areas are not used in accordance with their function, the permissible nighttime noise level during the nighttime does not apply.

[b] The downtown zone of cities with a population of more than 100,000 is the area of compact residential development with a concentration of administrative, commercial and service facilities. For cities with districts with a population of more than 100,000, a downtown zone may be designated in these neighborhoods if they are characterized by compact residential development with a concentration of administrative, commercial and service facilities.

(e.g., with a level of 58 dB), and the next 7 hours have lower sound levels (e.g., 54 dB), the average of these 8 hours (the so-called equivalent level) will still be below the 55 dB limit for single-family housing. Conversely, in the nighttime, the regulation indicates that the reference time is one least favorable hour of the nighttime. This means that out of the 8 hours of the night (22:00 to 06:00), the highest level of turbine operation from only 1 hour is taken, and this value is compared to the limits of 45 dB (for homestead, multi-family or commercial development) or 40 dB for single-family development.

The main problem during the registration of noise from wind farms is the selection of optimal parameters of wind speed driving the turbines, while maintaining the meteorological conditions specified by the applicable methodology (Regulation – emissions 2021). It is necessary to meet the condition of conducting measurements at an average wind speed at the control point (usually 4 m above ground level), not exceeding the value of 5 m/s. At the same time, when conducting measurements, one should strive for the wind speed, at the height of the axis of the rotors of wind turbines, to reach the values corresponding to the maximum acoustic power of the turbines or close to these values. In practice, meeting both of these is difficult and depends on the development of the area, the location of measurement points and the existing surrounding screening elements. However, it does happen that both conditions are met during measurements, i.e., the turbines operate at full power and the average wind speed does not exceed 5 m/s at a height of about 4 m at the measurement point.

In 99% of measured cases, most – but not all – of the wind turbines of the farm operate at maximum power. This is a common phenomenon. It is primarily due to the unpredictable variability of vertical wind speed profiles over time, as well as the terrain and land use around wind farms. The distribution of turbines on the farm also has a significant impact.

When conducting simulations (calculations) of noise from industrial sources (e.g., wind farms) on the basis of the regulation (Journal of Laws 2021 item 1710, Appendix 7) (Regulation – emissions 2021) the algorithm found in the PN-ISO 9613-2:2002 (PN-ISO-9613 2002) standard is used. The sound propagation simulation algorithm allows carrying out calculations with an accuracy of ± 3 dB (Evans & Cooper 2012; Probst et al. 2013; Zagubień & Ingielewicz 2017).

1.1.9   Results of Noise Measurements around Wind Farms

All the results of equivalent noise levels presented below were performed during post-implementation environmental monitoring from 2005 to 2022. The measurements were carried out by accredited testing laboratories with the participation of the authors of this monograph.

Wind turbines with a horizontal axis of rotation were tested. The acoustic power of the tested facilities was in the range of 101–107 dB and was achieved in the speed range of 10–13 m/s, at the height of the nacelle. The rated power of the turbines was in the range of 1.5–3.0 MW, and the length of the rotor blades ranged from 33 to 58 m. The height of the towers ranged from 78 to 160 m. Due to the lack of exceedances of permissible noise levels during the daytime (no case of such exceedances was ever found), the results for the nighttime were compiled. In accordance with the regulation on permissible noise levels in the environment (Regulation – noise 2007), the results refer to 1 hour of nighttime. Only measurement points located near residential buildings, up to about 500 m from the nearest wind turbine, were included in the compilation. These were the points located closest to the outermost turbines on the analyzed wind farms. There were between 1 and 11 such points around each of the farms summarized below. The results at control points more than 500 m away were not included in the compilation, since the noise levels (equivalent sound levels) recorded there are much lower than those presented; a similar relationship can be seen in the results presented by other authors (Pawlaczyk-Łuszczyńska et al. 2018) (Table 1.8).

TABLE 1.8
Results of Wind Farm Noise Measurements with Recorded Wind Speeds During the Study – Nighttime

Farm No.	Number of Turbines	Output of a Single Turbine	Tower Height	Equivalent Nighttime Sound Level L_{AeqN}	Average Wind Speed at the Height of the Turbine Nacelles	Average Wind Speeds at Measurement Points at 3.5 m Height
	(Units)	(MW)	(m)	(dB)		(m/s)
WF1	1	2.0	78	37.9–39.2	-	0.9–1.2
WF2	3	2.0	79	42.1–42.3	-	3.1–3.2
WF3	1	1.6	80	37.2–38.9	8.9	3.8–5.0
WF4	8	1.5	80	36.9–39.8	8.8	3.2–4.6
WF5	14	2.0	80	40.6–42.2	8.1	3.0–4.5
WF6	10	2.0	80	36.4–41.0	-	3.0–4.1
WF7	30	3.0	90	41.6–43.2	8.5	3.1–4.7
WF8	22	2.0	95	41.3–43.8	7.9	3.0–4.5
WF9	16	2.0	95	39.7–42.4	8.0	3.1–4.6
WF10	23	2.3	98	35.5–44.3	6.7	1.8–2.3
WF11	1	2.0	100	37.2–40.9	8.4	3.2–4.5
WF12	35	2.0	100	38.0–41.7	8.9	1.8–3.6
WF13	12	2.0	100	45.1–46.3	8.9	3.0–3.2
WF14	15	2.5	100	39.5–42.8	7.9	3.0–4.5
WF15	16	2.5	100	39.2–41.8	8.8	3.1–4.7
WF16	6	2.5	100	37.5–40.6	7.6	3.0–4.3
WF17	11	2.0	105	36.1–44.7	7.2	1.8–4.5
WF18	24	2.0	105	41.7–44.3	10.5	0.4–2.6
WF19	27	2.0	105	36.7–39.4	10.6	2.2–3.6
WF20	18	2.0	105	36.6–44.1	11.3	2.7–4.5
WF21	1	3.0	105	39.3	8.5	2.7
WF22	3	2.0	105	34.7–37.3	7.4	1.5–4.2
WF23	11	2.3	115	37.3–40.4	7.5	2.7–3.6
WF24	18	2.3	115	41.5–44.9	9.9	1.3–4.5
WF25	5	2.5	120	41.9–42.5	8.2	3.3–4.7
WF26	11	3.0	120	41.8–42.5	9.3	3.9–4.8
WF27	10	3.0	120	43.6–44.8	10.2	3.7–5.0
WF28	2	2.5	160	41.3–44.2	-	3.5–5.0

At a distance of 500 m, among the 28 farms surveyed (about 100 measurement sessions), the noise limit (45 dB) was exceeded only once. Owing to the possibility of performing silencing operations (software changes of "mods"), this power plant currently does not cause exceedances of noise limits in the environment. On most installations, measurements were performed several times, at different times of the year (spring, summer, autumn, winter). The areas around wind farms are dominated by homesteads, for which the permissible level at night is 45 dB. In the measurement points analyzed, only four points, out of 27 wind farms, were representative of single-family residential development. Single-family housing development at nighttime is assigned a permissible level of 40 dB by the regulation (Regulation – noise 2007). The occurrence of single-family houses in the closest building line to the wind farm is extremely rare and does not exceed 1% of other types of development, for which a noise limit of 45 dB applies at nighttime. For reasons of confidentiality, details of farm locations are not provided. Only the results of measurements and general data, relevant to acoustic assessments, are presented to allow conclusions to be made.

Results of control measurements of infrasound noise around wind farms – during the measurements, efforts were made to follow the guidelines of the reference methods of environmental noise measurements in the audible range (Journal of Laws 2021 item 1710, as amended, Appendix 7) (Regulation – emissions 2021).

In Poland, there are no established permissible levels for infrasound noise and no reference methods describing how to measure infrasound noise in the environment. There are also no permissible levels of infrasound noise in the working environment, which were withdrawn in 2009. The withdrawal was related to the significant levels of infrasound noise recorded in long-distance road transportation vehicles, exceeding the 102 dB level that was in effect until 2009 as a permissible level. Due to the lack of symptoms in drivers, 102 dB was considered too low to be in effect as a permissible level. To date, a new limit value has not been set, which appears to be necessary due to the continued inspection of workplaces for infrasound noise.

The Danish Environmental Protection Agency (DEPA) recommends that environmental infrasound exposure levels be 10 dB below infrasound hearing thresholds. According to the publication (Jacobsen 2001), the G-corrected hearing threshold for people with special sensitivity is 95 dB. DEPA recommends that the average total level corrected by the G-frequency characteristics in the frequency range up to 20 Hz should not exceed 95 dB. When G-frequency-corrected sound pressure levels are considered, then ISO 7196 (PN-ISO-7196 2002) or DEPA indicates that values lower than 85 or 90 dB will always be below the perception or annoyance thresholds. According to Danish standards, the permissible level of infrasound (including from wind turbines), in residential areas, inside living rooms, classrooms and offices, is set at 85 dB (G), both during daytime and nighttime.

The only legal act currently in force in Poland, in the field of infrasound, is the PN-Z 01338:2010 standard (PN-Z-01338 2010), which defines the principles of measurement and the so-called criterion of nuisance of infrasound noise at workplaces. Equivalent levels, which constitute the criterion of annoyance, are 102 dB (86 dB for jobs requiring special concentration of attention), with reference to 8 working hours or a working week, with G-curve-corrected measurements (PN-ISO-7196 2002). Although these values do not provide a basis for assessing infrasound noise hazards in the environment, they were referred to in summary. As a result, it was shown that infrasound noise levels around wind farms are relatively low. The obtained measurement results around wind farms were also compared with infrasound levels commonly found in the environment (Table 1.9).

The measured equivalent levels of infrasound (L_{Geq}) around wind farms do not exceed 102 dB (Pawlaczyk-Łuszczyńska et al. 2018; Bilski 2012; Ingielewicz & Zagubień 2014, 2013; Maijala et al. 2020), defined as the reference level in PN-Z-01338 2010. At the locations 500 m away from the wind farm, the level does not exceed 86 dB (reference level for workplaces requiring special attention). Examples of infrasound values from natural sources are summarized in Table 1.10.

TABLE 1.9

Results of infrasound noise measurements of wind farms along with recorded wind speeds during the tests

WF	Number of Turbines	Output of a Single Turbine	Tower Height	Equivalent Nighttime Sound Level L_{Geq}	Average Wind Speed at the Height of the Turbine Nacelles	Average Wind Speeds at Measurement Points at 3.5 m Height	Measurement Location
	(szt.)	(MW)	(m)	(dB)	(m/s)		
1	2	3	4	5	6	7	8
WF1	3	2.0	80	77.9–80.2	8.0–9.7	2.9–4.3	500 m from the farm
WF2	3	2.5	100	77.7–79.7	6.8–8.8	2.1–4.1	700 m from the farm
WF3	25	2.0	100	87.0–88.9	9.0–11.0	2.8–4.5	At the turbine tower
WF3	25	2.0	100	66.2–67.6	8.9–11.0	2.8–4.5	500 m from the farm

TABLE 1.10

Results of Infrasound Noise Measurements from Natural Sources

Type of Source	Equivalent Sound Level L_{Geq}	Average Wind Speed at a Height of 10 m	Average Wind Speeds at Measurement Points at a Height of 3.5 m	Measurement Location
	(dB)	(m/s)		
1	2	3	4	5
Noise caused by the forest	72.2–87.8 / 59.1	9.5–10.5	2.8–4.6	10–20 m from the edge of the forest in a building 25 m from the edge of the forest
Noise caused by the waves of the sea	76.1–89.1 / 64.3	9.5–11.0	2.9–4.8	10–20 m from the seashore in a building 25 m from the seashore

The results of the measurements obtained allowed concluding that infrasound noise from wind turbines is comparable to the noise levels measured for typical natural noise sources. The latter, to a greater or lesser extent, always accompany the operation of wind turbines, and in such cases they constitute an acoustic background, impossible to eliminate during turbine noise measurements.

An interesting analysis of infrasound noise emitted during daily household and living activities is presented in the articles by Zagubień and Wolniewicz (2016, 2017). The study site was located tens of kilometers from the wind farm. It was shown that daily exposure to infrasound noise ranges between 77.3 and 94.5 dB. These levels are comparable to those recorded outside a building at distances of 500 m or more from the wind turbines.

In the end, given the measurements carried out and the analysis of their results, it can be concluded that, in terms of infrasound noise, the noise from trees found in the vicinity of buildings plays a greater role than the noise generated by wind turbines. The observed levels of infrasound noise from wind turbines are lower or comparable to the noise associated with typical natural sources of infrasound commonly found in nature and infrasound noise accompanying humans in daily living activities. The measured levels of infrasound noise from wind turbines do not exceed those specified in the Danish criteria.

As far as the sounds emitted by wind turbines are concerned, the vast majority of scientists agree that there is no evidence that the noise from wind turbines, including the infrasound, has a negative impact on human health or well-being. The problem of infrasound effects of wind turbines is discussed extensively in a collective work (Massachusetts Department of Environmental Protection 2012), where it was shown that there is no basis for concluding that infrasound noise from wind turbines has a negative effect on humans.

1.1.10   Noise Maps and Simulation Analysis of Determining Noise Ranges of Wind Turbine Operation

Noise forecasts performed for wind farms aim at determining noise levels under worst-case conditions, which are usually defined as conditions in which the recipient is directly upwind from the wind turbine. In practice, it is rare for a recipient to be directly "upwind" of all the turbines that make up a wind farm (Evans & Cooper 2012). A wind turbine during its normal operation does not generate a constant noise level; it mainly depends on the strength of the wind at the height of the nacelle. In order to reliably represent the acoustic climate prevailing at the wind farm and reflect noise levels at residential developments, using noise simulation, a number of assumptions must be made. In order to reliably demonstrate the modeling principles for determining noise ranges from turbine operation, the least favorable variant was assumed, which means:

a. operation of all turbines at full sound power: in practice, this situation never occurs due to local wind power losses between neighboring turbines, which further translates into lower rotational speed of some turbines and lower noise levels they generate,

b. no influence of the background noise: in the case of wind turbines, these are household sounds, mainly involving the operation of other equipment in the

area of the residential building under study, such as the operation of agricultural machinery or traffic noise from nearby local roads or the sounds of
animate nature, having a significant impact on the measured noise levels,
especially during the summer months. The measurement experience of the
team compiling this study shows that sound levels of an animate nature,
e.g., crickets and birds or the noise of trees, can generate noise levels above
40 dB at night, which is close to the applicable limit values. Naturally, in the
winter months, this problem is less frequent and measurement evaluation of
wind turbine operation is much easier,

c. failure to take into account the directionality of wind turbines: a wind turbine does not generate noise evenly in every direction during operation.
This means that on the sides of the rotor, sound levels can be even 3–4 dB
lower than those in the direction perpendicular to the rotor plane (front and
rear). In the case where there will be four turbines around one residential
development at similar distances and in four different geographic directions, the noise emissions from the two side turbines will be lower than
from the other two – the one in front of the building and the one behind
the building (looking along the wind direction). At the acoustic modeling
stage, each wind turbine is assumed to be an omni-directional noise source
that generates a sound wave uniformly in each geographic direction. Thus,
taking each turbine in its acoustic modeling as an omni-directional source
generates a certain "noise impact fuse" compared to the field conditions for
which acoustic studies are performed. In addition, during the operation of a
wind farm there will be wind from any geographic direction.

Noise emission forecasting was carried out based on the algorithms of the PN-ISO-9613
2002 standard. It is in accordance with the recommendations of the current regulation (Regulation – emissions 2021), which indicates the abovementioned standard as
a method for calculating environmental noise ranges. The tool used for simulation
analyses was the CadnaA program from the German company DataKustik GmbH.

The basis for the analyses was a computational model including a prepared digital terrain model on which a single wind turbine was located. Control (reference)
points marking acoustically protected areas were determined around the turbine.
The model takes into account the absorbing acoustic properties of the terrain, as well
as its geometry, and sound propagation is favorably downwind.

An operating wind turbine was assumed in the simulation model as a point source,
along with acoustic parameters including its acoustic power level, determined from
actual field tests. The acoustic parameters of the operating turbine were determined
in accordance with the methodology contained in the international standard IEC
61400-11, which specifies how to determine the acoustic parameters of operating
wind turbines, including their sound power levels (IEC 61400-11 2012). Wind speeds
and corresponding acoustic power levels are given by turbine manufacturers in different ways. Most often, the relationship is presented in the technical documentation
in the form of a relationship between the wind force at the height of the nacelle and
the calculated acoustic level of turbine operation for that wind class. Many manufacturers also provide data on the level of acoustic power broken down into values in

octave and one-third octave bands, which makes it possible to carry out calculations of the noise range taking into account the influence of the ground using the general method, in accordance with the current standard (PN-ISO-9613-2 2002). In technical documentation of some turbines, the maximum sound power level is given for two wind speeds, i.e., at the level of 10 m and the corresponding wind speed at the level of the wind turbine rotor.

1.1.10.1 Assumptions of the Simulation Analysis

The specificity of wind turbine operation is associated with variable noise emissions, closely linked to meteorological conditions, and in particular to wind speed. Wind turbines begin operation above a threshold wind speed, which is usually 2–3 m/s (at the height of the nacelle), above which, as the wind speed increases, the rotor speed and the turbine generator system increase as well, and with them noise emissions. This increase is not linear and occurs only up to a certain wind speed. This speed is usually 7–10 m/s (at the height of the nacelle), and its exact value depends on the type and model of the turbine. Above this speed, the speed of the turbine rotor no longer increases, and the sound power of the device stabilizes.

Due to the high variability of wind turbine operation, which is difficult to predict, and thus variable noise emissions, the least favorable variant of the assessment was adopted for the purposes of this monograph, which assumed uniform operation and constant noise emissions from the wind turbine under study. In other words, it was assumed that the turbine's operation is characterized by constant and at the same time maximum noise emissions to the environment. Such a situation in reality is extremely rare, but it fulfills the so-called precautionary principle, i.e., the assumption of the least favorable acoustic situation with regard to the impact of noise on the environment. This is also in accordance with the current provisions of the Regulation of the Minister of Environment on permissible noise levels in the environment (Journal of Laws 2014 item 112) (Regulation – noise 2007).

The main objective of the analyses was to show noise ranges for turbines with different acoustic power levels, for different heights of wind turbine towers and taking into account typical sound propagation conditions.

Noise simulations were performed only for a single wind turbine, in order to clearly see the differences in distances for given wind turbine operating parameters. Any other approach to conducting noise simulations, such as calculations for more turbines or consideration of additional residential locations, would generate the need to analyze further noise simulation variants, and this would distort the noise calculations performed. The acoustic model for a single turbine allows understandable inference, which includes changes in the set parameters of its operation.

Noise simulations were performed for the following criteria:

1. Wind turbine parameters
   a. Nacelle height: currently installed – 130 m (together with the length of the blades, it can give a total height of the turbine equal to 190 m) and 180 m (including the blade, the total height may reach up to 250 m) – to be installed in the near future,
   b. Sound power level: 104, 105, 106 and 107.7 dB.

2. Ground absorption coefficient
   a. $G = 0$ (100% hard ground – e.g., a frozen lake in winter, asphalt roads, or paved yards – e.g., parking lots),
   b. $G = 1$ (100% absorbing ground – e.g., meadows and farmland in the off-winter period when the ground is frozen and icy; absorbing ground is all kinds of farmland found in agricultural and rural areas),
   c. $G = 0.5$ (50% hard ground and 50% absorbing ground – the conventional division between frozen water table and "soft" farmland) – in fact, the vast majority of the sites of currently operating wind farms consist of areas with a ground absorption coefficient close to the value of 1, since the amounts of hard surfaces – such as roads or paved yards – are hundredths of a percent of the area around the turbines,
   d. Without taking into account the A_{gr} soil factor – the so-called alternative method.
3. Noise limit values of 40 and 45 dB, for which noise ranges were determined.

1.1.10.2 Wind Turbines Adopted for Analysis

The noise calculations carried out to determine the distance between a wind turbine and a residential building (reference point) were performed for four sample turbines with actual sound power levels:

Turbine 1 $L_{WA} = 107.7$ dB with an electrical output of 2.2 MW
Turbine 2 $L_{WA} = 106.0$ dB with an electrical output of 3.2 MW
Turbine 3 $L_{WA} = 105.0$ dB with an electrical output of 2.0 MW
Turbine 4 $L_{WA} = 104.0$ dB with an electrical output of 3.4 MW

At this point, it should be mentioned that the A-curve weighted sound power level spectra of the turbines reflect the actual acoustic parameters of a given turbine, which were determined based on actual field noise measurements by the wind turbine manufacturer. The acoustic power level spectrum corresponds to the acoustic values determined for the frequencies of the middle octave or third bands (levels in dB for different Hz frequencies). The logarithmic sum of the values of the octave or one-third octave bands presents the summed sound power level spectrum, and the levels of individual octave or third bands are always lower than the summed value. Typically, the levels of individual bands in the spectrum are different and specific to a particular wind turbine.

Thus, there will be no rule stating that, e.g., if the total sound power level increases by, for example, 1 dB, the sound power level of the selected octave band in the spectrum will also always be higher. The shape of the spectrum is an individual characteristic of a given device and is affected by a number of variables. For example: for a turbine with a sound power of $L_{WA} = 104$ dB, the level in the 125 Hz, 4 kHz and 8 kHz octave bands is higher than for a turbine with a total sound power of $L_{WA} = 105$ dB, but in the other octave bands the $L_{WA} = 105$ dB turbine is louder than the $L_{WA} = 104$ dB turbine.

The spectra of the sound power levels of the wind turbines, which are corrected by the A-frequency characteristics and which were adopted for the simulation analyses, are shown in Figure 1.9.

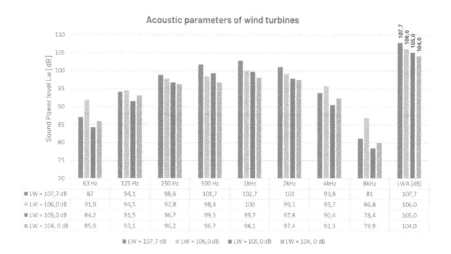

	63 Hz	125 Hz	250 Hz	500 Hz	1kHz	2kHz	4kHz	8kHz	LWA [dB]
▪ LW = 107,7 dB	87	94,1	98,8	101,7	102,7	101	93,8	81	107,7
▨ LW = 106,0 dB	91,9	94,5	97,8	98,4	100	99,1	95,7	86,8	106,0
▪ LW = 105,0 dB	84,2	91,5	96,7	99,3	99,7	97,8	90,4	78,4	105,0
▨ LW = 104, 0 dB	85,9	93,1	96,2	96,7	98,1	97,4	92,3	79,9	104,0

▪ LW = 107,7 dB   ▨ LW = 106,0 dB   ▪ LW = 105,0 dB   ▨ LW = 104, 0 dB

FIGURE 1.9   Acoustic parameters of the studied wind turbines.

The cited acoustic data for wind turbines represent the actual sound power levels of equipment that are currently being manufactured and installed on wind farms. It is estimated that the wind turbines produced in the future will have similar values of acoustic power levels as those currently installed on farms. This is due to two facts. First, the turbines will be increasingly higher, which means that the noise sources, i.e., the rotating blades, will be increasingly farther from the ground surface and thus farther from the point of noise reception. In practice, this means that the taller the turbine, the longer the sound wave has to travel and thus the more it attenuates (the acoustic wave weakens) in an elastic medium such as air. These will not be fundamental differences in sound levels, but it will be no such case that as the height of the turbines increases, the noise caused by their operation will be higher than for a smaller turbine height. Secondly, the length of the blades will become increasingly longer, which will probably translate into a lower rotational speed, if only because of the strength and load-bearing parameters of the blades. The overall noise caused by the operation of a wind turbine is mainly determined by the rotational speed. Thirdly, the use of serrated trailing edges on the blades (already in use today) will significantly reduce the aerodynamic noise of longer blades, which will also contribute to lesser noise from the next generation of wind turbines.

1.1.10.3 Noise Analysis Results

The purpose of the noise simulations was to determine noise ranges of 45 and 40 dB using different types and heights of wind turbines. The results of the noise simulations are presented in the form of tables and graphs containing the distances between the wind turbine and the residential building for:

- different sound propagation conditions above the ground ($G = 0$, $G = 0.5$ and $G = 1$),
- different permissible limits (40 and 45 dB),
- wind turbines with different levels of acoustic power.

Two main series of calculations were performed. For the currently used heights of wind turbines, the foundation of the nacelle was assumed at HH = 130 m above the ground (Hub Height) and the height of the turbine that will potentially be used in a few years, HH = 180 m above the ground (Table 1.11).

Conclusions for a turbine with a nacelle height of 130 m are as follows:

1. Operation of the loudest turbine adopted for noise simulation (sound power level equal to 107.7 dB), for all three considered ground coefficients and for the alternative method, will result in noise ranges of 45 dB below 500 m.
2. The farthest 45 dB range for the turbine L_{WA} = 107.7 dB is a distance of 425 m, for the least favorable ground absorption conditions ($G = 0$) and for the alternative method.
3. Maintaining a distance of 500 m for single-family residential development (40 dB at night) is possible for:
   a. turbine operation with a sound power level up to 104 dB, for a ground factor $G = 0$,
   b. turbine operation of L_{WA} = 104–106 dB, for ground factor $G = 0.5$ and $G = 1$.
4. The results above mean that the operation of the noisiest wind turbines adopted for simulation analyses, surrounded by homestead buildings, is within a distance of 500 m, without noise exceedances of 45 dB.
5. The operation of a single turbine with a sound power level of L_{WA} = 104 dB will not be louder than 40 dB in a distance buffer of 500 m, regardless of the noise propagation conditions at different times of the year (Table 1.12).

Conclusions for a turbine with a nacelle height of 180 m:

1. The operation of a single turbine with a sound power level of 107.7 dB, for all considered sound propagation conditions, will be less than 45 dB in a distance buffer of up to 500 m. In addition, the use of a turbine with a nacelle height of 180 m, for maximum sound absorption by the ground ($G = 1$), will not result in noise exceedances of 40 dB in a buffer of 500 m from the turbine.
2. Maintaining the 45 dB limit, for a turbine with a sound power level of L_{WA} = 107.7 dB, is possible well below 500 m. Depending on the propagation

TABLE 1.11

Isophone Ranges for Nacelle Height HH = 130 m

	Isophone Ranges (m)							
	$G = 0$		$G = 0.5$		$G = 1$		Alternative Method	
Sound Power Level (dB)	Isophone 45 dB	Isophone 40 dB	Isophone 45 dB	Isophone 40 dB	Isophone 45 dB	Isophone 40 dB	Isophone 45 dB	Isophone 40 dB
107.7	425	705	350	590	290	500	425	630
106.0	340	580	275	485	230	410	340	550
105.0	315	545	255	455	200	375	310	520
104.0	270	480	215	395	170	325	270	480

TABLE 1.12

Isophone Ranges for Nacelle Height HH = 180 m

	Isophone Ranges (m)							
	G = 0		G = 0.5		G = 1		Alternative Method	
Sound Power Level (dB)	Isophone 45 dB	Isophone 40 dB	Isophone 45 dB	Isophone 40 dB	Isophone 45 dB	Isophone 40 dB	Isophone 45 dB	Isophone 40 dB
107.7	400	690	325	580	260	485	405	680
106.0	315	570	250	470	190	390	315	570
105.0	290	530	225	435	160	355	290	535
104.0	240	465	175	375	115	300	240	465

conditions adopted, this distance can vary from 260 to 405 m, for absorbing ($G = 1$) and reflecting ($G = 0$) ground, respectively. This means, the operating noise of the loudest turbine will always be less than 45 dB in a buffer of 500 m from the turbine.

3. Turbine operation with noise less than 40 dB at night and in the buffer up to 500 m from the turbine is possible for turbine operation with sound power level up to 104 dB, for ground factor $G = 0$, and for turbine operation from $L_{WA} = 104–106$ dB, for ground factor $G = 0.5$ and $G = 1$.

4. For a turbine with a sound power level of $L_{WA} = 104$ dB, the noise is kept below 40 dB for all sound propagation conditions – the noise is always less than 40 dB in the buffer up to 500 m (Figures 1.10 and 1.11).

When comparing the impact ranges of turbines with nacelle heights of 130 and 180 m above the ground surface, the following conclusions can be drawn:

1. Noise ranges from turbines with taller support towers are smaller than those from lower turbines – provided the method is used to simulate noise, taking into account ground absorption and maintaining the same sound power levels of the turbines.

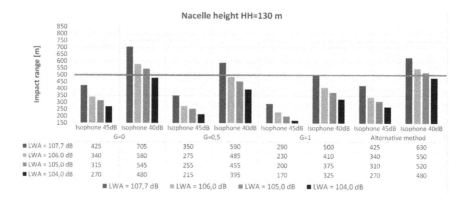

FIGURE 1.10 Impact range for nacelle height of 130 m.

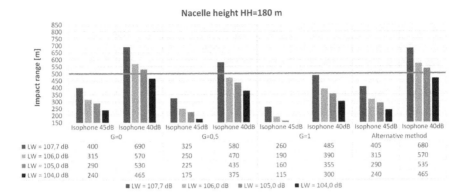

FIGURE 1.11   Impact range for nacelle height of 180 m.

2. Adopting limit values of 45 dB (nighttime limit for homestead, multi-family or commercial development) will not result in noise exceedances even for the most restrictive sound propagation conditions – noise at a distance of 500 m was not exceeded in any case.

3. The absence of noise exceedances of 40 dB (nighttime limit for single-family housing) for a distance of 500 m is achievable for wind turbines with a sound power level of no more than 104 dB. In practice, this means that if the nearest residential buildings function as single-family housing, then a turbine at a distance of 500 m cannot generate a sound power level greater than 104 dB.

4. For typical conditions for the location of wind turbines in Poland, such as cultivated fields and meadows (which, in terms of sound propagation over the ground, means assuming a ground factor of $G = 1$), the noise limit of 40 dB at a distance of 500 m will be preserved for the operation of wind turbines with sound power levels close to the value of 107.7 dB. In these areas, the 500-m distance limit without noise exceedances of 40 dB will be preserved even for the operation of the loudest turbines – with the sound power level close to 108 dB.

At this point, it should be clearly emphasized that the currently manufactured wind turbines have the technical ability to realistically reduce sound emissions during their operation. In other words, each turbine has systems for significant silencing of rotor operation, which can reduce noise at the "source" by up to 6 dB. Reduction of turbine operation noise is realized by:

a. Changes in blade angle, which allow reducing sound power levels by up to several decibels. Each turbine in its technical documentation has details of sound power level reduction for successive "mods" (operating modes). These are very common methods of reducing noise during turbine operation, which are implemented through software and which switch on automatically in situations of defined wind speed and direction at nacelle height. These are maintenance-free systems.

b. Special serrated trailing edges installed on the blades that minimize the generation of aerodynamic noise. Their effectiveness is currently at about 2 dB. Figure 1.12 shows examples of serrated rotor blade edges.

c. Repowering of wind turbines, which in acoustic terms means replacing existing turbines with new units that are smaller and quieter. The greatest advantage of repowering is to reduce the number of turbines within the operation of an existing wind farm, after a certain period of operation of the investment, for example, after 15 years. This is when some of the existing turbines are dismantled and replaced by more modern equipment. The new turbines will be more efficient, which will translate directly into their fewer numbers, thus reducing the noise range toward the nearest residential buildings.

In order to illustrate the effectiveness of the available methods of protection against turbine noise, a noise simulation was prepared for a farm consisting of ten wind turbines, for which noise ranges were reduced by applying the aforementioned noise reduction methods. The following analyses were conducted on the example of a specific wind farm location.

Variant 0 – the current one, includes the following parameters:

a. the total electrical capacity of the wind farm is 20 MW,
b. the number of turbines is ten, with a capacity of 2 MW each,
c. the sound power level of the turbine $L_{WA} = 106.4$ dB,
d. nacelle height of 100 m,
e. total height of 145 m (Figure 1.13).

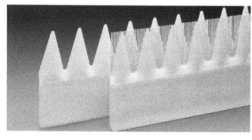

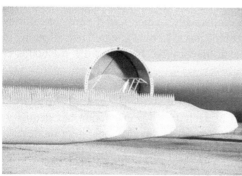

FIGURE 1.12 Examples of serrated trailing edges on turbine blades – PWEA materials.

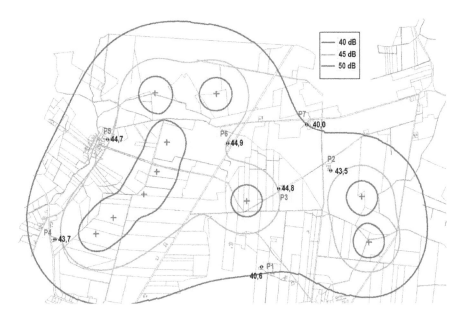

FIGURE 1.13   Variant 0 of noise analyses.

The obtained sound levels at the reference points are close to the 45 dB limit:

a. P3: 44.8 dB,
b. P6: 44.9 dB,
c. P5: 44.7 dB.

Variant 1 – number of wind turbines reduced to six units (use of repowering), with an electric power of 3.45 MW and a sound power level of 107.4 dB:

a. the total electrical capacity of the wind farm is 20.7 MW,
b. six turbines, with a capacity of 3.45 MW each,
c. acoustic power level of the turbine $L_{WA} = 107.4$ dB,
d. nacelle height of 132 m,
e. total height of 195 m (Figure 1.14).

Sound levels at reference points for Variant 1:

a. P3: 39.2 dB (44.8 dB before) – the turbine in the vicinity of P3 was removed, resulting in a substantial reduction in noise at P3,
b. P6: 42.7 dB (44.9 dB before),
c. P5: 42.0 dB (44.7 dB before).

Variant 2 – reducing the number of wind turbines (repowering) to six units with an electric power of 3.45 MW and a sound power level of 107.4 dB, and additionally equipping the rotor blades with silencing serrations (with effectiveness of 2 dB):

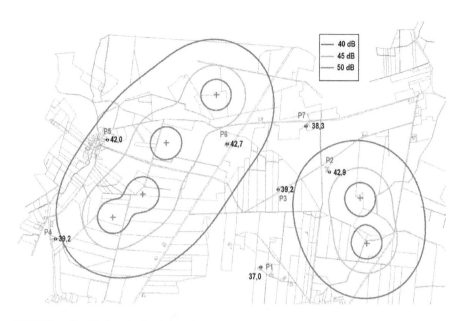

FIGURE 1.14   Variant 1 of noise analyses.

  a. total electrical power of the turbines 20.7 MW,
  b. six turbines with a capacity of 3.45 MW each,
  c. sound power level of the turbine $L_{WA} = 104.4$ dB,
  d. nacelle height of 132 m,
  e. total height of 195 m (Figure 1.15).

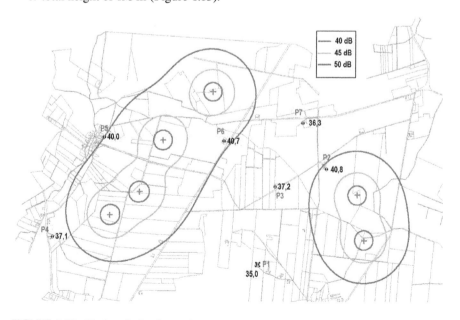

FIGURE 1.15   Variant 2 of noise analyses.

Sound levels at reference points for Variant 2:

a. P3: 37.2 dB (Variant 0–44.8 dB) – the turbine in the vicinity of P3 was removed, resulting in a substantial reduction in noise at P3,
b. P6: 40.7 dB (Variant 0–44.9 dB),
c. P5: 40.0 dB (Variant 0–44.7 dB before).

A summary of the difference in sound levels (Table 1.13), after reducing the number of turbines (repowering) and using turbine blade serrations (Figure 1.12), indicates that noise reduction of up to 7 dB is feasible.

In order to facilitating inferring and identifying the relationship between the number of wind turbines, their height and varying noise propagation conditions, noise changes are shown for the simplest case, i.e., the operation of a single turbine in the vicinity of a residential building. Such installations, i.e., single turbines located about 500 m away from residential buildings, are most common in Poland. It is natural for a wind farm in Poland to consist of a larger number of turbines, but the number of wind turbines within a radius of about 500 m from a residential building is usually no more than one or two. Thus, if it is assumed that one turbine is located within 500 m of a residential building, the examples shown above can be implemented directly to the planning of new wind farms. This applies both to simulated sound isolines ranges and to actual noise levels recorded during typical wind turbine operation.

In a situation where a second wind turbine is operating within a radius of 500 m from a residential development, then, according to the basic formulas for the summation of sound, the increase in noise will be a maximum of an additional 3 dB ($10 \log(n)$, where n denotes the number of wind turbines, so $10 \log(2) = 3$ dB). Naturally, the assumption above is a certain simplification, since the change in sound levels at residential buildings is affected by the specific distance of the turbines from the building. However, for the purposes of simple inference, it can be assumed that if there are two wind turbines within 500 m of a building, the calculated distances for sound power level $L_W = 105$ dB will be replaced by the approximately calculated distances determined for sound power level $L_W = 107.7$ dB.

TABLE 1.13
Summary of Results Obtained from the Analyzed Variants

Variant	Variant 0	Variant 1		Variant 2	
Reference Point	WO	W1	W0–W1	W2	W0–W2
			dB		
P1	40.6	37.0	−3.6	35.0	−5.6
P2	43.5	42.9	−0.6	40.8	−2.7
P3	44.8	39.2	−5.6	37.2	−7.6
P4	43.7	39.2	−4.5	37.1	−6.6
P5	44.7	42.0	−2.7	40.0	−4.7
P6	44.9	42.7	−2.2	40.7	−4.2
P7	40.0	38.3	−1.7	36.3	−3.7

However, it should be borne in mind that each wind farm project is different in terms of the planned turbines and the area where they will operate. That is why determination of specific sound levels for specific turbine placements in relation to residential buildings should be carried out on a case-by-case basis, using reference calculation methods to see how much the noise will increase for a specific distance, such as 500 m, with more turbines in the area.

To sum up the conducted simulation analyses, it can be concluded that:

a. The use of even the noisiest currently manufactured turbines will not cause noise exceedances for the limit values of 45 dB during the night (e.g., homestead development) at a distance of less than 500 m from the wind turbines.

b. For a larger number of wind turbines located within 500 m of a residential building with a noise limit of 40 dB, noise ranges should be determined based on noise calculations using dedicated simulation programs that comply with the recommendations of reference calculation methodologies.

c. If more than one turbine will be located within 500 m of a building, however, they must be equipped with viable silencing systems (e.g., changing the blade angle of attack and/or noise-reducing serrations).

d. The effectiveness of noise reduction and verification of noise emissions during modified turbine operation must always be preceded by certification measurements and described in the technical documentation of the equipment.

e. One should not use all the available options for turbine noise reduction at the stage of sound emission simulation. Some options should be left to apply emission adjustments resulting from control measurements of the actual wind farm.

1.2   OPTICAL IMPACT – SHADOW FLICKER AND LIGHT REFLECTIONS

The rotating rotor blades of a wind turbine cast a shadow on the surrounding areas, causing the so-called shadow flicker effect, also incorrectly called the stroboscopic effect. The shadow flicker effect is mainly observed during short periods of the day, i.e., in the morning and afternoon, when the low-lying sun in the sky shines from behind the turbine, and the shadows cast by the rotor blades are strongly elongated. The effect is especially noticeable in winter, when the angle of the sun's rays is relatively small (Figure 1.16).

This phenomenon, at frequencies above 2.5 Hz, is often called the stroboscopic effect. In the case of modern wind turbines, referring to this effect is wrong because, for a working turbine to achieve it, the rotor of the wind turbine would have to make 50 rotor revolutions per minute. Meanwhile, modern slow-speed turbines rotate at a maximum speed of 20 revolutions per minute, and the flicker frequency does not exceed 1 Hz (NCBiR 2021).

A literature review (Koppen et al. 2017) shows that not all countries have guidelines or legal regulations limiting and evaluating the so-called shadow flicker effect. Many countries that have such regulations rely on the German guidelines "Hinweise zur Ermittlung und Beurteilung der optischen Immissionen von Windenergieanlagen." The countries that do not have regulations for the optical impact in question often use the German guidelines as best practice.

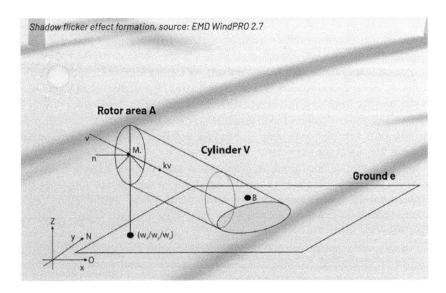

Shadow flicker effect formation, source: EMD WindPRO 2.7

FIGURE 1.16   Shadow flicker effect formation (Nawrotek 2012).

German guidelines limit the value of the shadow flicker effect at 30 hours/year and 30 minutes/day for the worst possible scenario, i.e., cloudless skies all year round. For individual cases, the time of exposure to this effect must in turn be limited to 8 hours/year. In turn, Table 1.14 shows the guidelines for other countries.

Moreover, in Germany, an analysis was carried out for a point 500 m away from a 110-m-high wind turbine. The shading effect at this point occurred only between January 12th and February 6th as well as from November 7th to December 2nd.

TABLE 1.14
Limit Values for the Shadow Flicker Effect

Country	Assumed Limit	Type of Regulation
USA	30 hours/year and 30 minutes/day	Guidelines
United Kingdom	No limits – but 30 hours/year and 30 minutes/day is often adopted	Guidelines and common practice
Japan	30 hours/year	Common practice
Brazil	30 hours/year and 30 minutes/day	Common practice
Canada	30 hours/year and 30 minutes/day	Common practice
India	30 hours/year and 30 minutes/day	Common practice
Australia	30 hours/year	Guidelines
Austria	30 hours/year and 30 minutes/day	Common practice
Belgium – Flemish region	8 hours/year and 30 minutes/day	Legal act
Belgium – Walloon region	30 hours/year and 30 minutes/day	Legal act
Norway	30 hours/year and 30 minutes/day for a worst-case scenario; 8 hours/year for a real scenario	Guidelines
Italy	No limits	None

A total of 11 hours of flicker were recorded during this period. These effects were recorded in the afternoon (3:00–4:00 pm) and never lasted more than 30 minutes during the day (Kowalczyk 2014).

In Poland, there are neither legal standards nor guidelines for maximum human exposure to the so-called shadow flicker effect. In this regard, there is only some common practice, the limit of which is set at 30 hours/year (GEO Renewables S.A. 2014).

Figure 1.17 shows changes in the size and shape of the shadow flicker effect area for wind turbines in relation to latitude. The visible differences are due to the variable position of the sun in the sky relative to the earth.

Simulation studies conducted by I. Piasecka show that the highest level of the shadow flicker effect – from 1,000 to 30 hours/year, is recorded at a distance of up to 500 m from the studied wind turbine, which confirms the legitimacy of locating residential buildings at a distance of not less than 500 m from the wind power installation (Piasecka 2014). This is also confirmed by Figure 1.18.

To assess compliance with the recommended limits, the shadow flicker effect should be modeled and predicted based on an astronomical worst-case scenario, which is defined as follows:

- the sun shines continuously and the sky is consistently cloudless from sunrise to sunset;
- there is sufficient wind to continuously rotate the turbine blades;
- the rotor is perpendicular to the direction of the sun's rays;
- sun angles less than 3° above the horizon are not considered (due to the possible occurrence of vegetation and buildings);
- distances between the rotor plane and the tower axis are negligible;
- refraction of light in the atmosphere is not taken into account.

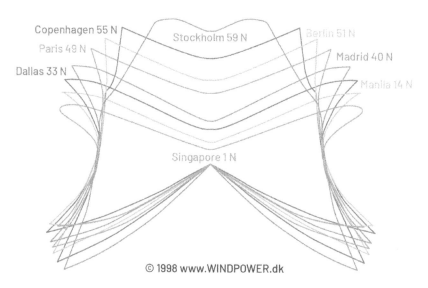

© 1998 www.WINDPOWER.dk

FIGURE 1.17   Shape of the shadow flicker effect area in relation to latitude (Danish Wind Industry Association 2003).

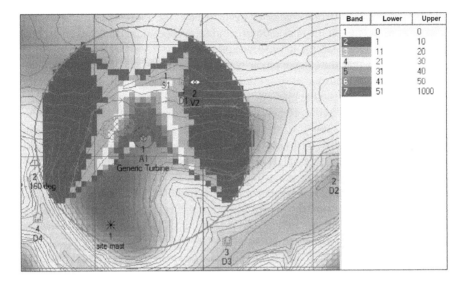

Band	Lower	Upper
1	0	0
2	1	10
3	11	20
4	21	30
5	31	40
6	41	50
7	51	1000

FIGURE 1.18   Approximate values of the shadow flicker effect (hours/year) caused by a 2 MW wind farm; grid: 1,000 m (Piasecka 2014).

Another optical phenomenon associated with the presence of a wind turbine in an area can be light reflections, resulting from the reflection of light rays from the turbine components. Currently, this effect is practically fully mitigated by the use of special paints and surfaces that reduce the reflection of light rays, and no significant nuisances are found in this regard.

Scientists confirm the annoyance of optical impacts on humans, primarily to people suffering from photosensitive epilepsy, who are particularly sensitive to light and flicker frequencies of 5–30 Hz (Fortin et al. 2013). However, as explained earlier, the frequency for wind turbines does not exceed 1 Hz.

Fortin et al. (2013) pointed out that shadow flicker from modern wind turbines is not harmful even to the people diagnosed with epilepsy, because the theoretical frequency of flashes from blade rotation at which an epileptic attack could occur would have to be 3–5 flashes per second (180 rpm), and the blades of large wind turbines cannot rotate that fast at normal operating speeds of 10–20 rpm.

Moreover, according to the British Epilepsy Association (British Epilepsy Society 2020), for shadow flicker phenomena to be a potential problem for the people with photosensitive epilepsy, which affects approx. 3% of the population diagnosed with epilepsy, several factors must be present simultaneously:

- the turbine blades would have to rotate at a speed greater than 3 Hz,
- the sun would have to be bright enough and at the right position and angle from the horizon relative to the turbine to cast shadows of sufficient intensity and length, which in practice does not often occur,
- a person suffering from photosensitive epilepsy would have to look at the turbine, and the sun should be behind the turbine.

In reality, the possibility of a combined occurrence of the above elements is low, and it can be considered that the operation of wind farms does not pose a health problem for people with diagnosed photosensitive epilepsy.

It is assumed that the phenomenon is not noticeable at all at a distance of ten times the length of the rotor blades. For objects closer, it is assumed that the setting of 30 hours of exposure to the phenomenon of shadow flicker per year is not harmful to human health (CPV Ashley Renewable Energy Company 2010). Studies conducted by Haac et al. (2022) show that already at a distance of 500 m from a single wind turbine and in cumulative terms, exposure to the effect of so-called shadow flicker does not exceed the time interval determined to be harmful to humans (Figure 1.19).

Simultaneously, light reflections, also known as the "disco effect," occur during sunny days when rotating rotor blades reflect the incident light. Light reflections are produced, which can disturb the field of vision. However, it should be pointed out that the use of matte paint on the rotor blades makes it possible to effectively reduce the optical impact (Baranowski et al. 2014).

Thus, it should be pointed out that there are no grounds for attributing the negative impact of optical impacts on human health to wind energy.

1.3 WIND TURBINES AND ELECTROMAGNETIC FIELDS

Electromagnetic fields accompany all equipment that produces, transmits or consumes electricity. Thus, a wind turbine, by its very nature, also generates such a field and has an electromagnetic effect on the surrounding environment.

Currently in Poland, such impacts are regulated by: Regulation of the Minister of Health on permissible levels of electromagnetic fields in the environment (Journal of Laws 2019 item 2448) and Regulation of the Minister of Climate on ways to verify compliance with the levels of electromagnetic fields in the environment (Journal of Laws 2020 item 258), which are implementing acts to the Environmental Protection Law (Journal of Laws 2020 item 1219, consolidated text). Associated with the Law on Environmental Protection are other legal acts that regulate the qualification, notification and determination of potential environmental impact and its monitoring (Regulation – electromagnetic fields 2007; Regulation – electromagnetic

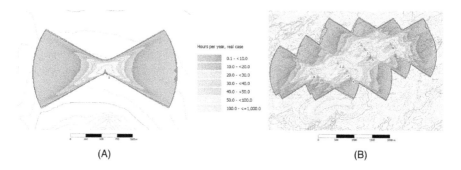

(A) (B)

FIGURE 1.19 Example of annual exposure to the shadow flicker effect around a single wind turbine (left) and cumulatively (right) (Haac et al. 2022).

fields 2010; Regulation – installations 2019; Regulation – electromagnetic fields 2003; Act – Information on Environmental Protection 2008).

The electromagnetic field is distinguished by its continuity of distribution in space, its ability to propagate in a vacuum and its forceful action on electrically charged particles of matter. The basic quantities that characterize the electromagnetic field include: f – frequency of the field [Hz], E – intensity of the electric component [V/m, kV/m], H – intensity of the magnetic component [A/m]. Sources of electromagnetic field, occurring in the environment, can be divided into two types: natural and artificial. Natural sources of electromagnetic fields include the natural radiation of the earth, the sun and the ionosphere. Of all the natural fields, the geomagnetic field is the best known. The intensity of this field ranges from 16 to 56 A/m. Above the earth's surface, there is also a natural electric field with an intensity of about 0.120 kV/m in normal weather. These are permanent fields. Of particular interest, due to their ubiquity, are artificial sources of electromagnetic fields at 50 Hz, mainly electrical equipment and machinery. The peculiarities of the electromagnetic field produced by such devices mean that the electric and magnetic components can be considered separately in its case. The magnetic field accompanies any current flow, while the electric field occurs wherever voltage appears. Typical magnetic and electric field strengths occurring in the vicinity of common-use devices are shown in Table 1.15, while Table 1.16 shows the frequency ranges of electromagnetic radiation and their areas of application.

The most common use of electromagnetic waves is in the telecommunications industry, where they are used as a carrier of information, hence propagation through space is also a very important issue. Electromagnetic waves are subject to all wave phenomena, i.e., reflection, diffraction or refraction. Therefore, the presence of

---

TABLE 1.15

Typical Magnetic and Electric Field Strengths, Occurring in the Vicinity of Common Equipment

Magnetic Field Values of 50 Hz Frequency Occurring in the Human Environment

Appliance	Magnetic Field Intensity
Automatic washing machine	0.3 A/m at a distance of 30 cm
Iron	0.2 A/m at a distance of 30 cm
Computer monitor	0.1 A/m at a distance of 10 cm
Vacuum cleaner	5 A/m at a distance of 30 cm
Shaver	12–1,200 A/m at a distance of 5 cm
Hair dryer	4 A/m at a distance of 10 cm

Electrical Frequency Field Values 50 Hz Occurring in the Human Environment

Appliance	Electric Field Intensity
Automatic washing machine	0.13 kV/m at a distance of 30 cm
Iron	0.12 kV/m at a distance of 30 cm
Computer monitor	0.2 kV/m at a distance of 10 cm
Vacuum cleaner	0.13 kV/m at a distance of 30 cm
Shaver	0.7 kV/m at a distance of 5 cm
Hair dryer	0.8 kV/m at a distance of 10 cm

TABLE 1.16

Frequency Ranges of Electromagnetic Radiation and Their Areas of Application

Frequency	Application
0–300 Hz (SELF, ELF)	DC electric traction, electrostatic technology, DC transmission lines, 50 Hz electric traction, electric power, communications
0.3–3 kHz (ULF)	Acoustic frequency control, medicine, communications, induction furnaces, quenching, brazing, melting, refining
3–30 kHz (VLF)	Telecommunications, radio navigation, medicine, induction heating, brazing, melting, quenching, refining, screen monitors
30–300 kHz (LF)	Radionavigation, marine and aeronautical telecommunications, carrier energy telephony, radiolocation, screen monitors, induction melting of metals, impedance tomography, ultraviolet, ignition systems
0.3–3 MHz (MF)	Telecommunications, radio navigation, amateur radio, AM radio, RF welding, packaging welders, medical
3–30 MHz (HF)	Frequency band for general use, radio modeling, international telecommunications, diathermy, magnetic resonance, dielectric heating
30–300 MHz (VHF)	Police, firefighting, amateur FM radio, VHF television, diathermy, ambulance, air traffic control, magnetic resonance
0.3–3 GHz (UHF)	Amateur radio, cab, firefighting, radar, radionavigation, UHF TV, microwave ovens, mobile telephony, diathermy, accelerators
3–30 GHz (SHF)	Radars, satellite telecommunications, amateur radio, firefighting, taxi, aircraft weather radar, police, radiolinks, burglar alarms
30–300 GHz (EHF)	Radars, satellite telecommunications, radiolines, radio navigation, amateur radio

various barriers in the environment, whether natural, resulting from the terrain, or artificial, created as a result of human activity, is important from the point of view of electromagnetic wave propagation. The permissible values of the physical parameters of electromagnetic fields are specified in the Regulation of the Minister of Health of December 17, 2019, on the permissible levels of electromagnetic fields in the environment (Regulation – electromagnetic fields 2019). This regulation differentiates permissible levels of electromagnetic fields for:

- areas intended for residential development,
- places accessible to the public.

Table 1.17 shows the range of frequencies of electromagnetic fields, for which physical parameters characterizing the impact of electromagnetic fields on the environment are determined for areas intended for residential development and permissible levels of electromagnetic fields, characterized by permissible values of physical parameters, for areas intended for residential development. The previously applicable regulations gave the same critical values for slow-moving electromagnetic fields, with the electric field in areas for human habitation (but without development) set at 10 kV/m, and this value is still applicable.

TABLE 1.17

Frequency Range of Electromagnetic Fields (Regulation of the Minister of Health of December 17, 2019)

Electromagnetic Field Frequency	Physical Parameter	Electrical Component	Magnetic Component	Power Density
No.	1	2	3	4
1	50 Hz	1 kV/m	60 A/m	-

The construction of wind farms results in the appearance of several potential types of electromagnetic field sources in the environment. These include:

- wind turbine generator,
- turbine generator transformer,
- cables inside the tower,
- underground cable network,
- a transformer station with instrumentation,
  - underground or overhead high-voltage line, leading energy from the farm's transformer/switching station to the point of reception by the public operator.

The main sources of the electromagnetic field, directly related to the wind farm, are the wind turbine generator and its transformer. These components are located inside the wind turbine nacelle at the top of the tower, i.e., at a height of about 100 m. All components of the power plant operate at a low voltage of 600–700 V. Medium voltage of 30 kV, which is the voltage of the farm's cable network, appears only at the output of the transformer. Due to the location of the wind turbine, as mentioned above, at a height of about 100 m, the intensity of the electromagnetic field generated by the elements of the power plant at ground level (at a height of 1.8 m) is negligible in practice. This also applies to turbines with lower towers. In the case of designed wind farm equipment, they are equipped with generators of relatively low power (generally 3–4 MW at present), compared to commercial turbogenerators (several hundred megawatts). These devices are mounted inside the nacelle, i.e., at a considerable height, hence their impact on the level of the electromagnetic field, measured at ground level (at a height of 1.8 m), is small, if measurable at all. It should also be noted that the equipment is located inside the nacelle and is enclosed in a space surrounded by a metal conductor with shielding properties, with the result that the effective impact of the wind turbine, on the "shape of the electromagnetic climate of the environment," will be zero.

Assuming considerable simplifications that do not include the shielding role of the nacelle housing, it is possible to approximate the level of electromagnetic field strength generated by the power plant components. The field generated by the generator will have a frequency of 50 Hz. In the case of a wind turbine with a height of about

100 m, the incident electric field strength at a height of 1.8 m will be about 9 V/m, i.e., well below the naturally occurring value. The resultant magnetic field will be about 4.5 A/m at this location, which is also less than the natural magnetic field. For turbines with higher towers, the values will be even lower. If turbines with lower towers are used, the electromagnetic field strength may be slightly higher, but will still be at a level much lower than the permissible values for areas accessible to the public. The research of the Central Institute for Occupational Safety (Gryz & Karpowicz 2016) covered in detail the wind turbine generator (in the immediate distance) and the space right next to the wind turbine tower. Short-term exposures with quite high values of magnetic field strength, relevant for the employees of companies servicing the turbines, but not exceeding safety requirements, were recorded – see Figure 1.20. It is also worth noting that in the electrical systems of wind turbines there are

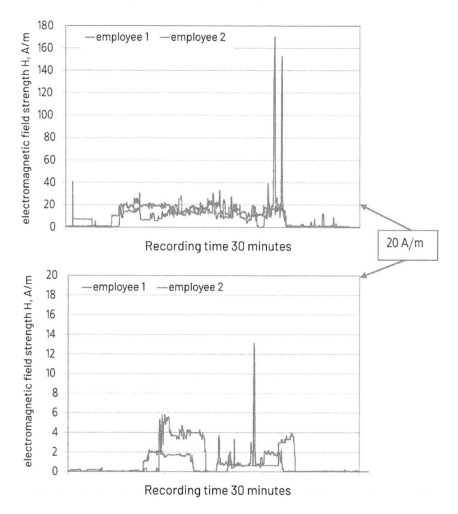

FIGURE 1.20   Magnetic field strength inside the wind turbine nacelle and in its vicinity according to (Gryz & Karpowicz 2016) observed for two employees.

AC/DC converters, which synchronize the variable frequency of the output voltage of the generators with the frequency of the power grid (50 Hz). Their basic components are insulated-gate bipolar transistors (IGBT) operating in pulse modulation mode. The operating frequency of such circuits reaches up to several tens of megahertz (MHz). Through a system of shielding and filtration, the radiation impact of the electromagnetic field generated by these devices and extending beyond the space of the nacelle is minimized, as confirmed by tests of the Central Institute for Occupational Safety (Gryz & Karpowicz 2016). In addition, these devices, in accordance with Directive 2014/30/EU, undergo stringent electromagnetic compatibility tests in accredited laboratories and present certificates confirming this fact before entering the market.

Thus, it should be concluded that the wind farm, together with its basic technical infrastructure, is not a source of electromagnetic radiation (i.e., emission of electromagnetic waves of medium and high frequencies) recorded near the ground surface. The source of such impact outside the turbines may be teletransmission transmitting antennas, used for the control and operation of the power plant. Such devices are characterized by low transmitter power and directional antenna radiation characteristics and do not pose a threat to the environment, especially since they are installed on top of power plant towers. In the case of cable (fiber-optic) links, which are most often used to control the operation of individual turbines, the use of sources of electromagnetic radiation of medium and high frequencies is completely eliminated.

The electricity supplied from wind farms is sometimes described as "dirty." This is justified by the high content of higher harmonics that distort the 50 Hz voltage sine wave, rather than its origin. This is not a justified view. According to the regulations contained in the IRiESD, 2020 and earlier editions, commissioning of wind farms includes tests and measurements, which should identify and eliminate such phenomena (originating from power electronic converters).

In conclusion, it should be said that wind turbines equipped with electrical power equipment, operating at 50 Hz (ELF – *extra low frequency*), do not pose a threat to the environment, as the fields they emit are many times lower than the values of field intensities allowed by Polish regulations (Regulation of the Minister of Health of December 17, 2019).

The second source of the 50 Hz electromagnetic field associated with the power system of wind farms are cable power lines. Their task is to deliver the energy produced at the wind farms to the substation and to the grid operator. Inside the area of the farms, medium-voltage cable lines are planned. These are the lines most commonly used in the Polish power system. The power grid cables are laid in trenches, in accordance with the standards in force in this regard. Together with the cables, a data communications fiber-optic network is also laid, which does not constitute a source of any electromagnetic radiation. Medium-voltage cable lines generate electromagnetic fields, the level of which is so low that they do not threaten the environment in any way. Only high-voltage lines above 110 kV are capable of generating electromagnetic fields with levels that may violate the electromagnetic climate quality standards. In the case of typical medium-voltage lines, electric field strength levels reach up to 0.6 kV/m. Typical magnetic field strength, on the other hand, does not exceed 4–5 A/m (with a standard of 60 A/m). The computationally determined distribution of the electromagnetic field around the 30 kV cable line, which carries a current of 250 A (equivalent to four 3 MW wind turbines operating at full power) is shown in Figure 1.21. As can be seen from the calculations, the electric field strength near the ground will be about 2 kV/m above the cable line itself, while at a height of 1.8 m it will take on a value of about 0.9 kV/m. These are much lower than the permissible values set for areas accessible to the public. As for the magnetic field, the intensity above the ground itself should not exceed 7 A/m, while at a height of 1.8 m it should be below 3 A/m. These values are also much lower than those permitted in the areas accessible to the public. In particular, it should be noted that the planned wind farm cable network is located outside residential areas; hence, the presence of people in the vicinity of the power line route is incidental. The electric and magnetic field intensity decreases very quickly and at a distance of a few meters from the axis of the cable line has a value close to zero – see Figure 1.21. In conclusion, the medium-voltage power grid of wind farms does not affect the deterioration of the quality of the electromagnetic climate of the environment and does not threaten human health and life (Figure 1.22).

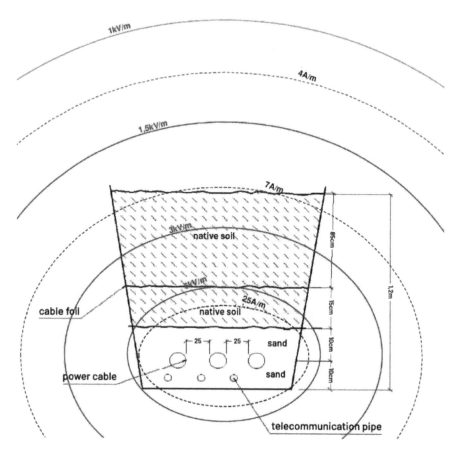

FIGURE 1.21 Electromagnetic field distribution over an exemplary 30 kV cable line transmitting a current of 250 A, (solid line indicates electric field isolines, dotted line – magnetic field isolines) according to (Stiller et al. 2006).

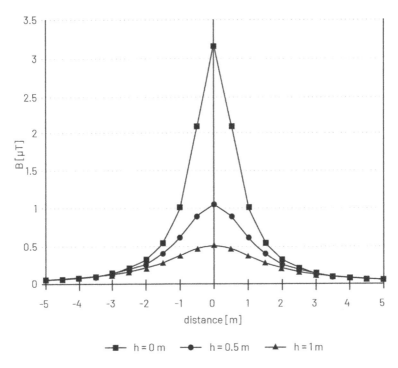

FIGURE 1.22 Magnetic induction distributions at a distance from the axis of a 20 kV cable line that transmits a current of 160 A (induction of 3.5 μT corresponds to a magnetic field strength of 2.8 A/m) according to Keikko et al. (1999).

Some publications and forums pessimistically forecast the creation of medium-voltage cable buses, by which the current from the entire farm will be supplied to the high-voltage substation; in the case of a 100 MW farm, this corresponds to a value of 2,500 A. This is not a realistic assumption, all the lines cannot be grouped in one bus, as this will negatively impact the reliability of the farm's operation and will be incompatible with the economics of grid operation. Power is always delivered to the transformer/switching station via several buses. Even if such a grouping of cable lines were to take place, it is worth noting that approaching the 60 A/m limit occurs only with six 220 and 400 kV cable lines, which transmit a current of more than 8,000 A (Figure 1.23) (Stiller et al. 2006).

Further transmission of power from the farm's transformer/switching station to the national grid, usually with a voltage of 110 kV, is carried out by means of an overhead or cable line. Its electromagnetic impact results from a regulation (Regulation – electromagnetic fields 2019; Regulation – electromagnetic fields 2007) and relevant standards, but it is not difficult to achieve the levels well below the requirements of the regulations. The only possible impact of wind farms on third-party property concerns the impact on radio wave transmission, i.e., on the reception of radio data communication signals or the reception of radio and television programs. Wireless communication systems use radio waves to transmit information between a transmitter and a receiver. In some cases, it is possible that the location of wind turbines

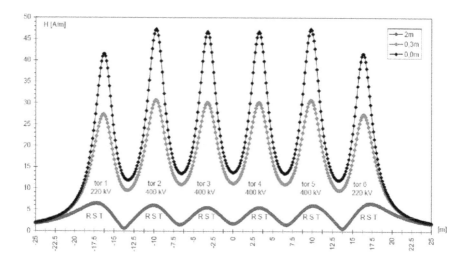

FIGURE 1.23 Magnetic field strength distributions at a distance from the axis of a six-track 220 kV and 400 kV cable line, which transmits a total current of 8200 A according to (Stiller et al. 2006).

could affect the reception of this information. Potentially, there could be four different mechanisms for a wind farm to affect the wireless transmission system (Karpat & Karpat 2016):

- electromagnetic interference – occurs when the electromagnetic radiation generated and emitted by wind turbines is contained in the band used by different services;
- near-field effect – occurs when wind turbines are located in the immediate vicinity of transmitters, and their operation changes the radiation characteristics of the transmitters;
- diffraction effect – occurs when the location of the wind park causes radio waves to be blocked on their way to the receiver, resulting in a decrease in signal strength;
- reflection effect – occurs when radio waves are reflected from the surface of wind turbines.

A detailed assessment of the abovementioned impacts is carried out by the State Radio Inspection on the basis of the submitted project documentation.

1.4   VIBRATIONS AND OSCILLATIONS FROM WIND TURBINES

A general definition of this phenomenon has already been cited in Section 1.1.2, using the example of "acoustic vibration." Vibrations of low frequencies, generally up to 100 Hz, propagating in solid media, are called vibrations or mechanical oscillations.

A wind turbine is a complex dynamic system consisting of jointly operating components. The operation of a wind turbine is based on the formation of a lifting force on the surface of the blade as a result of a stream of air flowing around it, causing rotor motion. Even assuming the uniformity of the wind flow acting on the rotor, it is necessary to bear in mind the effect of a temporary loss of lifting force associated with the passage of the blade near the turbine tower. Consequently, the support tower of a three-blade turbine will be excited by vibrations with a frequency three times the rotor speed. These vibrations, in turn, are transmitted to the foundation and then to the ground. Their further propagation in the ground can potentially be transmitted to buildings and the people in them.

The hazards caused by mechanical oscillations (vibrations) occurring in the environment, around wind turbines, are characterized below. When determining the harmfulness of vibrations to human health, the duration of their impact, the frequency of vibrations and their amplitude are evaluated. Particularly dangerous to humans are low-frequency vibrations. Every part of the body and organ are characterized by its own vibration frequency. If this frequency coincides with the frequency of vibration, resonance can occur, increasing the amplitude of vibration of the organ. The vibration's acceleration, expressed in m/s^2, or vibration velocity in m/s, is taken as the quantity that determines the amplitude of vibration.

In Poland, there are no specified levels of permissible vibrations in the environment. The evaluation of the harmfulness of vibrations is based on two standards (PN-B-02170 2016-12, PN-B-02171 2017). The standard PN-B-02170 2016-12 deals with the impact of vibrations transmitted through the ground to buildings, and caused by human activity. The sources of these vibrations may be located either within buildings (on separate foundations) or outside them. The scope of the standard includes rules for assessing the impact of the said vibrations, as well as rules for assessing their impact on vibration-sensitive apparatus located in buildings. The standard gives requirements for performing dynamic measurements to assess the impact of vibrations on buildings. Performing an assessment of the impact of vibrations on people, received passively, is made possible by the PN-B-02171:2017 standard. Passive reception occurs when a person is subjected to vibrations, when they do not operate the vibration source and do not directly affect the operation of this source.

According to the standard, the following vibration parameters can be determined:

- the RMS value of the acceleration or velocity of vibration, corrected in the frequency domain,
- the RMS value of the acceleration or velocity of vibrations, in one-third octave bands.

Assessment of the effects of vibrations on humans is carried out using two alternative methods: by measuring the corrected value of acceleration or vibration velocity over the entire frequency band or by measuring the spectrum of acceleration or vibration velocity in one-third octave bands. Measurements and evaluation are carried out in three mutually perpendicular x, y, z directions. Separate assessments are made for the horizontal x and y directions, which are the directions perpendicular to the human spine, and the vertical z direction along the spine.

The standard specifies the permissible values of vibration parameters that provide the required comfort, under various conditions of human habitation in residential premises. Vibrations in the frequency range of 1–80 Hz are evaluated. For each band of center frequency f from 1 to 80 Hz, the condition $a_{dop}(f) \leq n \cdot a(f)$ should be maintained, where a – the value of acceleration measured for individual center frequency f, a_{dop}– the permissible value of acceleration for individual center frequency f.

The permissible value for each one-third octave band is determined from the condition: $a_{dop}(f) \leq n \cdot a(f)$ where n – coefficient, for residential premises $n = 4$, at night, $a(f)$ – acceleration value for the human perception threshold according to the standard, for each frequency of the middle one-third f.

Table 1.18 summarizes the permissible vibration levels in residential buildings several hundred meters away from the wind turbines, determined as described above, as well as sample results of self-reported measurements at the wind turbine foundation. The recorded vibration levels at the point of contact between the wind turbine foundation and the ground were many times lower than the permissible levels applicable to residential buildings several hundred meters away from the wind turbine towers.

TABLE 1.18
Results of Vibration Measurements on the Foundation of a 2 MW Wind Turbine

Mid-Band Frequency f, (Hz)	Permissible Acceleration Value (m/s²) of Vibrations Perceived by a Human Dop (f)		Measured RMS Value of Acceleration (m/s²) of Vibration a(f)		
	Direction of Evaluation		Direction of Measurement		
	Z	X i Y	Z	X	Y
1.0	0.0400	0.014.3	0.0031	0.0036	0.0040
1.25	0.0356	0.0143	0.0034	0.0054	0.0036
1.6	0.0316	0.0143	0.0035	0.0035	0.0030
2.0	0.0283	0.0143	0.0032	0.0031	0.0080
2.5	0.0253	0.0178	0.0020	0.0022	0.0031
3.15	0.0228	0.0228	0.0021	0.0037	0.0035
4.0	0.0200	0.0286	0.0049	0.0051	0.0037
5.0	0.0200	0.0357	0.0044	0.0049	0.0074
6.3	0.0200	0.0456	0.0025	0.0043	0.0032
8.0	0.0200	0.0572	0.0051	0.0039	0.0036
10.0	0.0250	0.0716	0.0046	0.0048	0.0031
12.5	0.0312	0.0900	0.0029	0.0027	0.0034
16.0	0.0400	0.1144	0.0098	0.0070	0.0051
20.0	0.0500	0.1428	0.0051	0.0045	0.0039
25.0	0.0624	0.1784	0.0101	0.0030	0.0075
31.5	0.0789	0.2256	0.0037	0.0052	0.0078
40.0	0.1000	0.2856	0.0232	0.0092	0.0088
50.0	0.1252	0.3572	0.0556	0.0117	0.0158
63.0	0.1576	0.4520	0.0136	0.0021	0.0031
80.0	0.2000	0.5720	0.0178	0.0069	0.0034

In the course of authors' own research and in the literature descriptions, no case was found where the ground-borne vibrations from the operation of wind turbines reached values at the nearest residential buildings above human vibration perception. It is generally accepted that ground vibrations generated by wind turbines are so small that they cannot be felt by the people living more than 2 km from the nearest wind turbine (Nguyen et al. 2018). In fact, it is highly unlikely that the vibrations transmitted through the ground will be felt by people living more than 500 m from wind turbines.

There are a few scientific papers that suggest the ability of vibration waves transmitted through the ground to generate significant levels of infrasound (Gortsas et al. 2017). There are also researchers who dispute this (Nguyen et al. 2018) and suggest that the vibrations recorded on the windows of the building facade are well-correlated with the acoustic characteristics of wind turbines. Further studies of vibration levels in residential buildings in the vicinity of wind farms are warranted in order to determine why the described special cases occur.

1.5 MECHANICAL IMPACT – ICE PIECES AND BLADE FRAGMENTS

1.5.1 Description of Phenomena, Previous Observations and Incidents

Referring to the danger posed by the icing of wind turbine blades and the detachment of pieces of ice from them, it is necessary to analyze the mechanism of icing formation itself. During the formation of an ice layer on an airfoil, i.e., a rotor blade, environmental conditions are of particular importance. Most of the recorded events related to the detachment of ice chunks from wind turbine blades took place in the countries with a significantly higher number of days in which icing can occur than in Poland (Figure 1.24). In Poland, the number of days per year during which the weather conditions conducive to wind turbine rotor blade icing occur is less. According to the report (Tammelin et al. 1999), in Poland it is less than one day per year or 2–7 days per year in a smaller part of the country (Figure 1.24).

As far as the mechanism of ice formation on the surface of the blade is concerned, it should be noted that the most favorable conditions for the occurrence of this phenomenon take place when the turbine is stationary. During the operation of the turbine, icing forms to the greatest degree on the leading edge of the blade. In the remaining part, it has a layered form or does not occur, as can be seen in the photograph in Figure 1.25.

The paper by Hua et al. (2017) presents detailed results of studies pertaining to this process. They show that the maximum mass of ice ranges from 0.3 to 1.5 kg/m of the length of the blade and is greatest at its end. On the other hand, the thickness of the ice layer ranges from 1 to 5 cm and is also greatest at the end of the blade. This is shown in Figures 1.26 and 1.27. Similar results were presented in the paper by Ibrahim et al. (2018).

Therefore, it is impossible for heavy pieces (for example, 10 kg) to detach from the blade if a maximum of 1.5 kg of ice per meter is accumulated on it. This would mean an irregularly shaped piece, more than 6 m long with the drag coefficient CX much higher than for regular spherical or icicle-shaped pieces. This is, of course, a purely

FIGURE 1.24   Number of days in Europe when wind turbine icing occurs (Tammelin et al. 1999).

FIGURE 1.25   Icing of wind turbine blades (the photo on the right-hand side shows a turbine at standstill) (Baring-Gould et al. 2012).

theoretical consideration, since the published results describing events of ice pieces detaching from turbine blades show that they had lower weight.

The problem of the hazard posed by ice pieces detaching from wind turbine blades has long been the subject of evaluation and analysis. Due to the development of wind power in the countries with weather conditions conducive to formation of ice, a number of observations and experiments have been carried out for existing facilities.

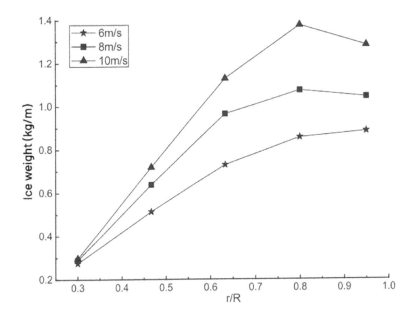

FIGURE 1.26 Distribution of ice weight along the length of the blade (Hua et al. 2017).

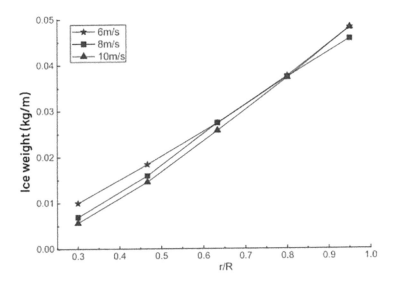

FIGURE 1.27 Distribution of the thickness of the ice layer along the length of the blade as a function of wind speed (Hua et al. 2017).

Their results can be found, among others, in the work of Cattin et al. (2014). This paper includes the results of measurements of the masses of ice pieces detached from turbine blades and their throw ranges calculated from the turbine tower. The research was carried out under extreme icing conditions (Swiss Alps and an area located at an altitude of 2,300 m above sea level). The test object was a 600 kW Enercon E40 turbine with no anti-icing or de-icing systems. The rotor diameter is 44 m. The research was carried out within the framework of the national research project "Alpine Test Site Gütsch," which concerned the measurement and prediction of icing on this type of structure. It was conducted by employees of a meteorological station located in the immediate vicinity of the turbine. In 2005/06 and 06/07, 121 ice pieces with a maximum length of 100 cm and a weight of up to 1.8 kg were recorded. The largest throw distance from the turbine was 92 m. However, such large distances account for only 5% of all events, as shown in Figure 1.28. In turn, Figure 1.29 presents the relative frequency of ice chunks of different weights detaching from the blades. The graph shows that the vast majority were pieces up to 0.2 kg. The pieces heavier than 0.5 kg constituted only 2%–3%. There was also no clear correlation between the mass of the ice piece and the distance of its fall from the turbine tower.

Similar results are presented in the paper by Renström (2015).

In the work of Bakoń (2013) the author notes that multi-annual statistics documenting accidents occurring in wind power industry recorded 34 cases in which

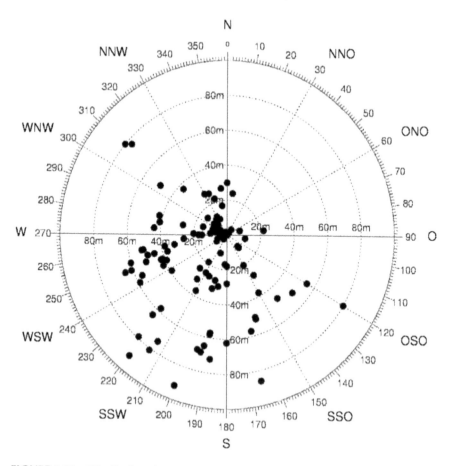

FIGURE 1.28 Distribution of detached ice pieces around the turbine (Cattin et al. 2014).

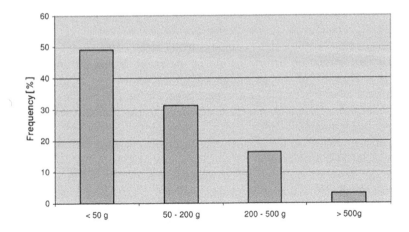

FIGURE 1.29 Relative frequency of detachment of ice pieces with different weight (Cattin et al. 2014).

the cause of damage was attributed to the detachment of ice pieces from the wind turbine. The farthest flight of an ice piece detached from a wind turbine blade was 140 m from the turbine tower.

Regardless of the views on the frequency of the phenomenon in question, the distance of the ice piece throw and the mass of the chunks, the problem defined in the literature as "ice throw" is a significant issue for the safe operation of wind farms. However, semantic issues should be noted here. In Poland, in colloquial language, the term "launching" is often used to refer to the pieces of ice falling off the blades. It is not appropriate, because a wind turbine is not a "launcher." The linguistic interpretation of the word "launch" means a powerful throw over a long distance. It cannot be considered a priori that the fall of an ice piece is such a launch (strong, over a long distance). Of course, the fact that the circular motion of the wind turbine blades gives these elements a certain initial velocity is relevant here. Translating the term directly from English, it can be considered that a wind turbine "throws ice," but it should not be explicitly stated that it "launches" it.

1.5.2 EVALUATION OF ICE THROW RANGE

For many years, determination of the maximum distance at which an ice piece can be thrown by a wind turbine has been the subject of analyses conducted both by specialists in theoretical mechanics and by specialists directly involved in the design and construction of wind turbines. This is because it is clearly related to the safety of people who may come within the range of such a throw.

In physical terms, the flight of an ice piece that falls off a wind turbine blade can be treated, in simplified terms, as the diagonal projection of a solid with a specified mass, initial velocity, aerodynamic drag and throw angle. The literature contains descriptions of the relevant equations of motion and their solutions leading to a simulation-determined trajectory of the throw and, most importantly, its range.

The following works can be mentioned as examples: Pojmański (2015), Bresden et al. (2017), Szasz et al. (2019), Renström (2015), Seifert et al. (2003), and Lennie et al. (2019). With the exception of the work of Pojmański (2015) the results of the simulation determination of the range of the ice chunks throw are similar. Despite the varied mathematical apparatus and assumptions, the maximum simulation-obtained throwing range for ice pieces does not exceed 350 m. Examples of simulation results obtained in the abovementioned works are shown in Figures 1.30–1.32.

The paper by Renström (2015) analyzed the process of ice detachment from the blades of a 3.3 MW turbine with a rotor diameter of 120 m. A ballistic projection model of the detached ice piece was developed. The maximum throw distance is 239 m with a blade circumferential velocity of 20 m/s at a radius of 55 m and a throw angle of 45°. After taking into account the wind speed for Swedish weather conditions, a maximum distance of 350 m was obtained. It should be mentioned that the author of the cited paper (Renström 2015) verified his results by studying an actual wind farm located in northern Sweden. The distances presented in the paper did not exceed 200 m, with a turbine with a mast height of 100 m being analyzed, and the assumed mass of the ice piece was 1 kg.

The most analytically advanced throw range calculations are presented in the paper by Szasz et al. (2019). They are characterized by calculations performed for 6 degrees of freedom. This means taking into account the rotation of a piece of ice detached from the blade. In these calculations as well, the determined throw ranges do not exceed a distance of 200 m, for turbines with a tower height of 100 m.

A common feature of all the works presented above is their invocation of Seifert's formula (Seifert et al. 2003), presented in 2003, defining the area free of thrown ice pieces as the outside of a circle with a radius expressed in meters:

$$R_s = 1.5(H + D) \tag{1.6}$$

Seifert's work originated as a summary of the Wind Energy Production in Cold Climate project developed in 1999 by the Finnish Meteorological Institute (Tammelin et al. 1999) (H is the distance from the ground surface to the rotor axis, D is the diameter of the turbine rotor). Therefore, for modern land-based turbines, the "safe" radius R_s is about 330 m. The formula indicated above was developed as a result of

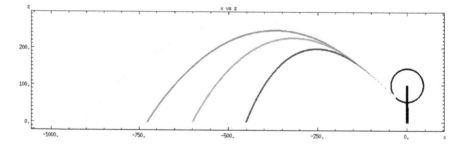

FIGURE 1.30 Trajectories of ice pieces with weight of 0.1 kg (black), 1 kg (grey), 10 kg (deep grey), theoretical range (light grey) according to the study (Pojmański 2015).

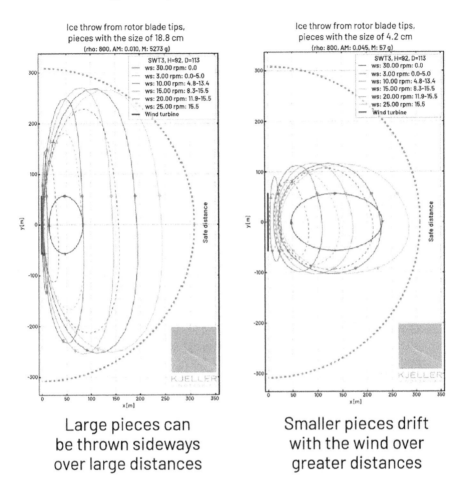

FIGURE 1.31 Ice throw ranges for different wind speeds (ranges are contained within a circle having a radius of 330 m) (Robinson et al. 2013).

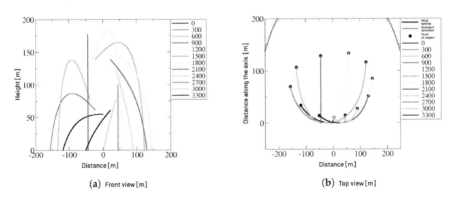

FIGURE 1.32 Flight trajectories of ice pieces weighing 1 kg, for different angles of detachment from the rotor blade, advanced simulation analysis (Szasz et al. 2019).

analytical work and empirical experience of the authors. However, they do not present the exact mathematical model used for the calculations. Characteristically, however, none of the works cited above (with the exception of Pojmański (2015)) disproves the results obtained by Seifert and casts doubt on the validity of the formula for determining the safe radius R_s.

It is also worth noting that the paper and report by Tammelin et al. (1999) cited above provide a formula for the range of ice piece throw when the wind turbine is stopped. It depends linearly on the wind speed and at 20 m/s reaches 200 m; thus, it is significant.

A significant opinion-forming role in the problems of mechanical impact of wind turbines in the case of blade icing, failure and fire, was played in 2012–2016 by the study of Prof. G. Pojmański (2015). It is easily found on the internet.

In the introduction to the study by Prof. Pojmański, there were a number of valid observations. They concerned the location of wind turbines in the "legal void," from the point of view of assessing their technical condition and controlling operational safety.

The fundamental theses of the study in question concern the extent of throwing of ice fragments and parts of turbine blades that have failed. As indicated in the considerations presented above, a number of articles and reports define the range of these throws at a distance of no more than 350 m, with the results reported being both simulation analyses and long-term observations of real objects (Seifert et al. 2003; Bresden et al. 2017; Renström 2015; Lennie & Pechlivanoglou 2019; Tammelin et al. 1999; Cattin et al. 2014; Carlsson 2011; Hua et al. 2017; Shohag et al. 2017; Szarsz et al. 2019).

The ice throw ranges in Prof. Pojmański's study exceed the distance indicated above by at least twice. The most general reason for obtaining such results are the assumptions made in the study under discussion and the overly simplified simulation model. Before proceeding to a more detailed presentation of the objections in this regard, comments should be made as to the research methodology used in the study as presented below.

- The author of that study presents the wind turbine as a mechanical system, the purpose of which is to launch pieces of ice and fragments of blades. Prof. Pojmański supported his considerations with the relevant ballistic vocabulary: "... any mass detaching from the tip of the blade becomes a projectile," "...can glide," ".... launching pieces of ice," "launched fragment," reinforcing the thesis of the extremely long distances and ranges reached by ice pieces and fragments of the turbine structure. The quoted phrases introduce a sense of danger in a reader who is unfamiliar with the principles of mechanics. Thus, the scientific study is presented on a prepared emotional "foundation," which is not the appropriate approach.
- It is widely believed that in engineering and science, the primary method of verifying theoretical hypotheses is to experimentally test their validity under real-world conditions. At the time of writing (probably 2012), many thousands of wind turbines with a total capacity of nearly 200,000 MW (NC RfG 2016) were already in operation in the world. As mentioned above, the problem of the ice throw range was verified by measurements

and observations (Tammelin et al. 1999; Cattin et al. 2014). Meanwhile, Prof. Pojmański refuted empirical inquiry, writing: *"It is often quoted that most of the launched pieces fall near the turbine. Simulations show that this is not true. Due to the peculiarities of the ballistic curve, a significant portion of accidentally launched pieces lands near the maximum range."* Considering the methodology of the research paper, this approach is astonishing – an empirical experiment involving a large research sample is negated because it does not correspond to the results of theoretical simulations on a simplified model.

- In a similar manner, Prof. Pojmański comments on the inconsistency of his results with actual observations. Referring to the remark that ice pieces were not found so far away (as indicated by the author), he states that *"…. they were probably not looked for there."* The comment of the authors of the study is similar to the point above.

- Theses put forward in scientific studies should be based on true assumptions. Unfortunately, in the discussed study, not all assumptions meet such a condition. The information that 90% of wind turbines installed in Poland come from German repowering is not true (and was not true a few years ago). According to the data from the Energy Regulatory Office provided to PWEA, it is currently (out of 6,000 MW) at most 15%. This number will soon decrease to zero, as used turbines are not eligible for auction proceedings, nor do they meet the technical requirements of the RfG regulation (NC RfG 2016). The blade failure rate of 10^{-2} quoted by the author is overestimated by dozens of times; according to the data shown in the following chapters, it actually approximates 10^{-4}. This difference is of fundamental importance in assessing the risk of accidents related to the destruction of turbine blades.

As indicated in Section 1.5, ice on a blade is subject to decay amounting to 1.5 kg/m. Thus, the most dangerous projection ranges given in Prof. Pojmański's study (a 10 kg piece – 550 m, a 10 kg icicle – 750 m) are determined for the objects that will not be formed in reality (such an icicle would be 6 m long). Obtaining a piece takeoff velocity v_0 equal to the blade tip velocity (e.g., 300 km/h) is impossible, since ice fall-off occurs mainly during wind turbine start-up, and heavily iced blades do not allow reaching rated speeds even in favorable winds. In order to give the ice pieces a linear speed equal to the tip of the blade, it would be necessary to construct a special mechanism to release the ice, as was done for siege machines in ancient and medieval times. The simplified way of taking into account the mechanism for releasing the ice pieces, as well as the air resistance and lifting force, may ultimately explain why the results obtained in the study under discussion do not correspond to the results and measurements presented in the literature indicated above. In view of Prof. Pojmański's denial of the reliability of observations of real objects, it would be appropriate to present, in particular, the work of researchers specializing in fluid mechanics (Szasz et al. 2019), published in 2019, reporting maximum throw ranges of 200 m and not denying Seifert's formula (Seifert et al. 2003).

1.6 MAJOR FAILURES, DISASTERS AND FIRES

The extreme case of mechanical damage to a wind turbine is a construction disaster involving the complete collapse or fracture of its tower structure – see Figure 1.33. Such cases are considered spectacular and rare, and are therefore subject to film documentation – https://www.youtube.com/watch?v=OXdXYPzSN7YW.

To some extent, they may be the consequence of a construction error and occur during the construction of the turbine. The probability of such disasters as a result of environmental effects (seismic shocks, temperature, groundwater) for new units built in accordance with the rules of construction art is low (CWIF 2020). According to the requirements of EN 1990: 2004/NA: 2010, wind turbines are built to ensure that the risk of disaster is less than 4×10^{-5}.

Significant danger results from the push of hurricane wind on the turbine tower and its blades. Failure of the blade braking system can lead to the turbine's rated speed being significantly exceeded, resulting in the breakage of the blade or part of it. Blade fracture and the resulting oscillation of force moments and excessively high stresses can lead to the collapse of the entire turbine structure.

Thus, blades are considered to be vital for the safe operation of a wind turbine. The blades of a turbine are elements that mediate the transfer of energy from the flowing working medium (wind) through the shaft to the generator. Blades, due to their important role in the functioning of a wind turbine, are among its most important

FIGURE 1.33 Wind turbine tower structure disaster – PWEA materials.

components. The operational safety of the entire system depends on their strength and durability. The destruction of blades is very often equivalent to the destruction of the entire turbine. The load on the blades of turbines with axial flow of the working medium can be systematized as follows:

- load due to the flow of the working medium (wind),
- mass load.

The first group of loads is caused by the static and dynamic action of the flowing medium on the blade surface. Mass loads arise from the rotational motion of the masses of the blades and associated components, as well as forces caused by elastic vibrations of the blades and the entire rotor. The mass (centrifugal) load, as well as the aerodynamic load of the blade, can be considered in three planes, which include bending and torsional loads. Flexural and torsional loads can occur because the center of aerodynamic thrust does not coincide with the axis of stiffness of the basic section blade.

During operation, the following stresses develop in the blades of axial turbines (Lipka 1967; Szczeciński 1982):

- tensile, caused by the centrifugal forces of the rotating blade masses,
- flexural from the action of the pressure of the medium on the profile part,
- flexural from the centrifugal forces of the rotating mass of the blade,
- tangential from the action of torsional moments of the forces caused by the action of the flowing medium,
- tangential from the action of torsional moments of the mass forces of the blade,
- flexural caused by transverse vibrations of the blade,
- tangential from torsional vibrations of the working part of the blade.

The effects of wind turbine blade fracture and its degradation process are shown in Figure 1.34.

A wind turbine blade has a certain length of the working part, which is under the action of centrifugal forces. Centrifugal force plays a very important role in shaping the trajectory of blade parts detached from a wind turbine rotor. In fact, a detached blade fragment does not take a trajectory tangent to the circle of radius on which it was located. An example of the flight path of detached turbine rotor blade fragments can be observed in the frames of the video footage showing a catastrophic failure – Figures 1.35 and 1.36 (the rotor rotation direction in both cases is clockwise). The pictures show the effect of aerodynamic forces (aerodynamic drag) during the tear-off of the blade (gradually released, as a result of the process of destruction of its structure and loss of cohesiveness of the material from which it is made) and the radial (under the influence of centrifugal force) scattering of elements falling under the influence of gravity. Blades, as a result of aerodynamic drag, decelerate rapidly once the damaged part is detached from one of them (they deflect in the direction opposite to the direction of rotation). When dealing with this type of issue, it is important to keep in mind the specificity of the turbine rotor (the rotor is not driven; conversely, it drives the generator by receiving kinetic energy from the wind), so it is

(a)

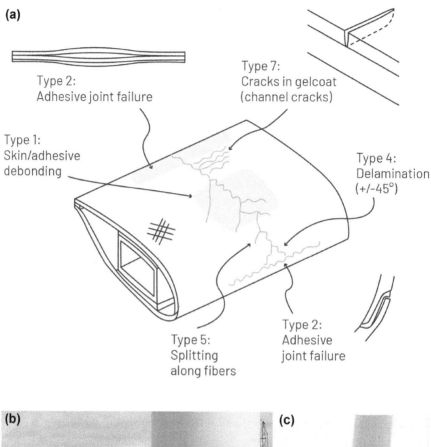

FIGURE 1.34 Damage to wind turbine rotor blades according to Shohag et al. (2017): (a) possible mechanisms of damage development, (b) broken off part of the blade at the Korsze wind farm, (c) progressive damage to the wind turbine blade (Poland, 2012, PWEA materials).

not able to maintain speed after the blade or blades are damaged. Thus, the destruction of the blades is associated with a sharp reduction in the speed of the wind turbine and the throwing range of the blade debris decreases.

FIGURE 1.35 A case of catastrophic failure of a turbine, the flight paths of blade fragments can be observed (parts of footage) (Wind Turbine Crash 150 km/h 2008).

FIGURE 1.36 A case of catastrophic failure of a turbine, the flight paths of blade fragments can be observed (parts of footage) (Wind turbine – EXPLOSION COMPILATION 2017).

The study by Professor Pojmański (2015) discussed in Section 1.5 assumes that *"In the case of a throw of a piece detached from a turbine, the boundary conditions are determined by the initial velocity v_0 and the point of detachment of the piece $x0$ (the initial velocity and position vectors of the piece), the wind speed, and the shape and mass distribution in the piece. The piece is affected by gravitational force Fg and aerodynamic force RA, the components of which – drag force RD and lift force RL – depend on the orientation of the piece with respect to the instantaneous velocity vector v."* In physics, the kinematics and dynamics of objects are very often considered under conditions when their motion is not perturbed. With such assumptions, formulas clearly describing this motion are obtained. Unfortunately, if one tries to apply these formulas in practice, the results of simulation calculations differ significantly from real results.

A commonly understood example would be the trajectory of a tennis ball or football. Trying to determine their trajectory based on mass, initial velocity and drag force does not correspond to the actual trajectory, for which the rotation imparted to the ball intentionally by the player as well as the effects of the wind are essential. The higher the player's skill, the more rotation is imparted to the ball in a more finely tuned manner, and the more attempts at analytical modeling of ball flight are complex and difficult.

Therefore, the issue of the detachment of a blade or its part (as well as the detachment of a piece of ice) cannot be treated as a classical oblique throw taking into account air resistance. The computational model presented in the study (Pojmański 2015) does not reflect the actual conditions taking place during a turbine failure. It is applied incorrectly and significantly deviates from the actual flight path of the analyzed component. Understandably, the calculations (correct in the sense of the mathematical description of the oblique throw) were carried out for extreme conditions. On the other hand, many factors contribute to the overestimation of even the extreme throw distances calculated in Pojmański (2015). The drag force coefficients of the "thrown" pieces were improperly taken into account. In the study, they are selected in a simplified way and do not take into account their actual shapes, which have much worse aerodynamic properties when it comes to the selected geometric models of the ice pieces. In order to obtain reliable results, it cannot be assumed that the detached pieces have the shape of a sphere or an icicle. Advanced tools that allow simulation of blade icing are now available. On their basis, the shape of the elements can be studied and their aerodynamic coefficients can be determined. An attempt at such advanced simulations was made in Szasz et al. (2019).

The instances of turbine rotor blade destruction documented in the videos clearly indicate the involvement of the centrifugal component in the movement, which is obvious given the very strong internal tension of the blades. Thus, the model adopted in the study discussed in terms of the predicted extent of damaged blade fragments yields highly overestimated results.

The throw range of blade fragments damaged as a result of failure of the turbine brake system (up to 2 km) or as a result of a fire with thermal support (ten times the height of the tower), as predicted by Prof. Pojmański (2015), is not confirmed in the available theoretical studies or in the documentation of emergency conditions of existing wind turbines.

Failure of the electrical system (usually a short circuit in the electrical circuits) or failure of the lubrication system for mechanical parts can lead to a fire located in the nacelle of the wind turbine, accompanied by the scattering of its burning fragments around the tower (Figure 1.37). Some such spectacular fires have been filmed – https://www.youtube.com/watch?v=H9liEPCgkc8. The fires on turbine blades result from lightning strikes, which occur despite the fact that the blades and tower are equipped with lightning protection systems. Fires from lightning occur at extremely high discharge energies, with an estimated 2% of discharges having such characteristics.

Turbine fires are most often located in their nacelle, and are difficult to extinguish, due to the lack of external access and the intensive air supply (even at standstill, it is significant at an altitude of more than 100 m). Improperly conducted maintenance or

FIGURE 1.37 Wind turbine fire according to PWEA (2015).

repair work inside the turbine nacelle can also be a cause of fire. Burning fragments of the nacelle and blades detaching from them can cause fires in the area around the tower, and the smoke from burning turbine parts contains toxic components.

The actions of the fire service, which does not have the capacity to intervene at such high altitudes, were reduced to containing secondary fires and securing the area from human access. However, nowadays, in situations where waiting for the structure to burn completely is not an environmentally acceptable option, a firefighting helicopter is involved.

1.7 PUBLIC ACCEPTANCE

Wind farms are projects meeting with a significant public reaction and public discussion regarding their location (Gamboa & Munda 2007). This is primarily related to the characteristics of these investments, which include almost 200 m tall structures, constituting a significant dominant landscape (Badora 2017). However, it should be borne in mind that the visual perception of wind turbines is purely subjective. Thus, the degree of impact of wind energy facilities on the aesthetic qualities of the landscape is difficult to measure (Niecikowski & Kistowski 2008).

Research on public acceptance of wind power is growing rapidly, but there is still a lack of knowledge about the different types of acceptance, whether acceptance is correlated with demographics, and what influences the acceptance of wind farms in the urban landscape (Westerlund 2020). Lack of public acceptance results in delays, public protests, increased costs and sometimes blockage of wind power projects (Bolwig et al. 2020), while increasing the risk of not achieving environmental policy goals (Cohen et al. 2014). Therefore, energy developers and policymakers need to understand public acceptance to ensure successful planning, implementation and operation of wind energy systems (Landeta-Manzano et al. 2018).

Despite widespread public support for wind power as a clean and low-cost technology, wind projects are often challenged by local communities. According to a 30-year study by Rand and Hoen (2017) public acceptance is now widely seen by wind energy practitioners as a significant barrier to the implementation of renewable energy (Mroczek 2011). According to the data, the reasons for public objection to wind power are noise, impact on human health, decrease in the value of parcels of land or negative impact on the landscape and general attractiveness of the area. Thus, it can be concluded that the visual impact of wind energy projects is often considered one of the main factors causing community concern. M. Wolsink points out (2007) that the relationship between visual impact and public acceptance is probably one of the most complex to understand, with many different points of understanding on this issue.

Public acceptance toward the impact of projects on the landscape is particularly influenced by the number of turbines, distances and aesthetic quality of these facilities (Ellis & Ferraro 2016; Betakova et al. 2015). Greater sensitivity to wind projects occurs toward wind projects in landscapes with high aesthetic quality and a higher level of acceptance in unattractive landscapes. Wind turbines are also better accepted if the structures are away from settlements, transportation infrastructure and vantage points (Molnarova et al. 2012).

However, the issue of the visual impact of wind turbines is not a new topic, as it was noted more than 20 years ago that the visual effects of wind turbine locations were one of the most common arguments raised by local residents against them (Krohn & Damborg 1999). It is indicated that greater attention to environmental impacts can increase public acceptance of wind turbines, ensuring their optimal location and ultimate contribution to reducing greenhouse gas emissions (Peri et al. 2020).

According to a study by Freiberg et al. (2019), the direct and indirect visibility of wind turbines can affect the health of residents, and reactions can vary when combined with noise. Moreover, the annoyance associated with the visibility of wind turbines may mediate between visual exposure and the health of residents. However, as confirmed by the author, further high-quality studies are needed to confirm these results.

Moreover, the combined prevalence of high annoyance from altered views and shadow flicker was 6% each from a total of 17 studies. Results on other health effects were inconsistent, with some indications that direct visibility of wind turbines increases sleep disturbance. The annoyance of direct visibility, flickering shadows and flashing lights was significantly associated with an increased risk of sleep disorders. Only one study indicated that reactions toward visual features of wind turbines may be influenced by acoustic exposure.

An interesting study was conducted by Szychowska et al. (2018), which aimed to investigate the influence of audiovisual information on the evaluation of wind turbine noise annoyance. The results showed that the sound level of the auditory presentation was the most influential factor in the evaluation of annoyance. The second most influential factor was the visual presentation sample, and the least influential factor was the audio presentation sample. At the same time, this research shows that the sound and audiovisual samples of wind turbines were evaluated similarly to the transport samples.

In contrast, other surveys show that there is an apparent split in public perception of wind farms. On the one hand, there is a need to develop renewable energy sources, and on the other hand there are serious concerns about the visual impact of wind turbines used for energy production (Molnarova et al. 2012). Fear of visual impact is a major factor influencing the public's reaction to the development of new wind farms. The study conducted by Molnarov et al. aimed to objectify this impact and identify the factors that determine how people evaluate these structures. They studied the visual quality of the landscapes in which these structures are to be placed, the number of structures and their distance from the viewer, and various characteristics of the respondents. It was found that wind turbines are also better accepted if their number in the landscape is limited and if the structures are far away from settlements, transportation infrastructure and vantage points.

Sklenicka and Zouhar (2018) proposed a method to objectify the assessment of the visual impact of wind farms. This method generates predictions based on landscape indicators. It allows skipping the stage of classifying landscape types, where the use of a subjective element is usually unavoidable. Simultaneously, this method makes the assessment more sophisticated, accurate and controlled. The objectivity of this method was further supported by conducting a sociological survey on a representative (demographically stratified) sample of respondents, each of whom filled in a questionnaire on the impact of WT on their visual preferences. This method for the first time uses map-based landscape indicators in a panoramic simulation to predict the visual impact of a wind turbine. It provides a better match between visual preferences and landscape index analysis than the mapping-based projections used to date. The method of objectified prediction of the impact of onshore wind farms makes it possible to automatically analyze the visual impact of wind turbines, both on small areas of interest and on a large region. The method provides a suitable basis for both preventive and causal forms of assessing the visual impact of wind turbines, and provides important support for objectifying the planning and decision-making process.

As reported by Knopper and Ollson (2011) the peer-reviewed studies found that wind turbine annoyance is statistically related to wind turbine noise, but found to be more strongly related to visual impact, attitude toward wind turbines and noise sensitivity. So far, no peer-reviewed articles have shown a direct cause-and-effect relationship between the people living near modern wind turbines, the noise they emit and the resulting physiological health effects. The reported health effects are likely attributable to a number of environmental stressors that cause a state of irritability/stress in some of the population.

Similarly, Pedersen and Larsman (2009) found that the people living in the areas where wind turbines are perceived to contrast with their surroundings (flat areas) were more likely to be annoyed by the noise than those living in hilly areas, regardless of sound pressure levels, if they thought wind turbines were ugly, unnatural devices that had a negative impact on the landscape. The increased negative reaction, according to the researchers, may be related to an aesthetic response, rather than to the authoritative effects of simultaneous auditory and visual stimulation.

The subjective assessment of residents' well-being can be influenced by the visual effect, attitudes toward wind farms and other subjective factors that vary over time (Pohl et al. 2021). While confirming the relationship between wind turbines and their impact on human health, most studies emphasize that this relationship is modeled by many variables, most of which are not physical in nature. The most common finding is that the presence of wind turbines can cause irritability. However, there is no conclusive literature data confirming the actual visual impact on human health.

Individual characteristics such as noise sensitivity, privacy issues and social acceptance, benefits and attitudes, local situation and wind farm planning conditions also play a role in assessing the impact of wind turbines on human health (Radun et al. 2022).

Figure 1.38 shows that the plans to build wind turbines or their actual presence in an area can lead to disruption and unrest among the local community, but the influence of the (planned) turbines can be shaped by many factors. Personal factors include attitudes, expectations and noise sensitivity. Situational factors include other possible influences, such as visibility or shadow flicker, other sound sources and type of terrain. Contextual factors include participation, decision-making process, location procedure and procedural fairness.

However, it has not been conclusively proven that the proximity of a wind turbine negatively affects human health, including triggering stress reactions, impairing quality of life and sleep. Ideally, preventive measures of sound levels across the frequency range and routinely collected health data from records could be used in conjunction with more subjective data (Kamp & Berg 2021).

The study by Huebner et al. (2019) found that the stress associated with noise annoyance was negatively correlated with perceptions of, among other things, a lack of reliability in the wind project planning and development process. Objective indicators such as distance from the nearest turbine and sound pressure levels modeled for each respondent were not found to be correlated with noise annoyance.

According to the WHO, wind energy is associated with fewer negative health impacts than other forms of traditional power generation, and will even have positive health effects by reducing emissions. Moreover, there is no reliable scientific evidence of the possibility of health consequences, interference or moderate vibrations from infrasound or low-frequency sounds generated by wind farms (World Health Organization 2018). *The people living near wind turbines are only exposed to certain noise annoyance, which should not be equated with a threat to human health. Noise annoyance, in turn, can lead to sleep disorders and psychological stress. However, the direct effect of noise emissions from operating wind turbines on sleep disorders or psychological stress has not been demonstrated* (Bakker et al. 2012).

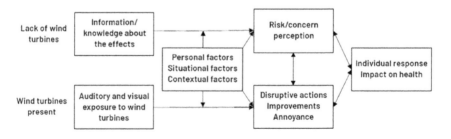

FIGURE 1.38 Graphical model summarizing the relationship between exposure to wind turbines and individual response (Michaud et al. 2016).

2 Methodology for Determining the Maximum Intensity of Negative Impacts of a Wind Farm on Human Health

2.1 ACOUSTIC IMPACTS

2.1.1 PERMISSIBLE LEVELS OF AUDIBLE NOISE

The noise generated by wind turbines has its own audible characteristics, such as the predominance of low-frequency sounds. Phenomena such as amplitude modulation as well as the impulsive and tonal nature of emissions may occur (Pedersen et al. 2009). In this chapter, the evidence that each noise source, including each wind turbine, has its own specific acoustic characteristics is cited (Zagubień & Wolniewicz 2019). However, each noise source should be verified for tonality, impulsivity and amplitude modulation, which can cause nuisance. Nevertheless, the phenomenon of tonality is rare, mainly during the operation of older types of turbines, and is easily eliminated through the use of special serrations on the blades. It should be borne in mind that the mentioned nuisances at distances above 500 m from wind turbines are practically imperceptible (Zagubień 2018). The adverse effects of wind turbine noise on human health can be divided into three groups: subjective effects (including annoyance and dissatisfaction), interference with activities (such as speech, sleep and learning) and physiological effects (such as anxiety, tinnitus or hearing loss), while one of the main human reactions to audible sounds is irritation (Pedersen et al. 2007). Other than this annoyance, scientific studies have found no direct negative health effects associated with wind turbine noise. It has been shown that sleep problems and feelings of discomfort may be a secondary effect of noise exposure, which has been linked to noise annoyance (Pedersen et al. 2007). The same conclusion was reached by Rakun et al. (2022), who also indicated that the noise level generated by wind turbines was related only to annoyance, not to the harmfulness of the emissions on humans. Moreover, the literature does not indicate that infrasound or low-frequency sounds cause different effects than higher-frequency sounds (Van Kamp & van den Berg 2021).

This emphasizes that sound emissions from wind turbines do not cause hearing loss (Council of Canadian Academies 2015). It is also important to point out that human perception of sound depends mainly on background noise in the environment (Biswas 2021).

For a modern turbine, the maximum sound power level is in the range of 100–110 dB(A). For a listener on the ground near the turbine, the sound level outside will be no higher than about 55 dB(A). In residential areas, the level is often lower, and most studies have shown that few, if any, people are exposed to average sound levels above 45 dB(A).

In the case of a wind turbine, the maximum sound levels recorded in residential areas are not much higher than the average environmental sound levels (Van Kamp & van den Berg 2021).

The assessment of noise pollution is carried out on the basis of the Regulation (Journal of Laws 2014 item 112) (Regulation – noise 2007), which defines permissible noise levels for different types of sources. Wind turbines are not separate noise sources, so they are included in the category of "Other facilities and activities that are a source of noise." The noise limit value of a wind farm for one least favorable hour of the nighttime is 40 or 45 dB, depending on the land use. In comparison, road noise has a permissible level of 56 dB, with this value averaged over 8 hours. Thus, it can be seen that in the worst hour of the night, the permissible level of road noise is about 60 dB, so the difference in permissible levels of road noise and wind turbines is about 15 and 20 dB, respectively. None of the noise annoyance studies conducted justify such a difference between the permissible noise levels from industrial sources versus traffic sources. In addition, wind turbine noise increases with wind speed, resulting in higher background noise levels. At wind speeds of about 10 m/s, the background noise level associated with the noise of the wind itself is about 55 dB, which is greater than the nighttime limit by 10 and 15 dB, respectively. Thus, in this context, the permissible values of noise from wind turbines should be considered appropriately low, i.e., guaranteeing acoustic comfort in acoustically protected areas.

It is difficult to agree with the statements presented by Barbara Lebiedowska in "Farma wiatrowa jako sąsiad społeczności wiejskiej. Oddziaływanie akustyczne farm wiatrowych" ("Wind farm as a neighbour of a rural community. Acoustic impact of wind farms"), according to which there are no established permissible levels of noise from wind facilities, as well as the suggestion to adopt a permissible level, corrected by the C (dB(C)) characteristic curve. To assess the acoustic impact of wind farms, the same method is used as to determine the impact of the operation of industrial plants. Such an action should be considered correct, since a wind farm is undoubtedly an industrial installation (Zagubień 2018). This methodology, also in the case of industrial plants, does not take into account amplitude modulation or infrasound noise. Importantly, it cannot be unequivocally assumed that a possible new assessment methodology will allow more accurate analyses of the acoustic climate around wind farms. Current sound propagation simulation algorithms allow conducting calculations with an accuracy of ± 3 dB (Biswas 2021; Zagubień & Ingielewicz 2017).

It should be pointed out that the sound level expressed in dB(C) is used to identify a possible high level of low-frequency noise, as the difference between the equivalent sound level C and A (Zagubień & Wolniewicz 2020). Another example of the use of the C correction curve is the measurement of peak sounds at workplaces, the permissible level of which is 135 dB(C) (Hojan 2014). This is a very high level, not found in such intensity in the vicinity of wind turbines. In contrast, acceptable A-curve-corrected sound levels are commonly determined around the world.

The author's proposal to perform measurements using C or LIN correction, rather than A correction, as is currently the case, is probably aimed at including low-frequency and infrasound components in the measurements. Here, it should be emphasized that the A correction curve, which attenuates the lower frequencies of the audible range, was created specifically to adapt sound pressure measurements to the perception of sound by the human ear (Kirpluk 2017). Thus, it should not be omitted when measuring noise in the audible range of wind turbines.

It should also not be overlooked that Prof. Barbara Lebiedowska, when assessing the impact of wind turbine noise, does not cite the works that would unequivocally demonstrate the negative acoustic impact of wind turbines on human health. Moreover, in her studies, Prof. Lebiedowska makes use of literature items without affiliations, whereas their authors do not have academic degrees. Thus, in the opinion of the authors of this study, the questioned text of Prof. Lebiedowska lacks significant research value. It should be considered one-sided, presenting only the arguments of wind energy opponents.

2.1.2 MINIMUM DISTANCES OF WIND FARMS FROM RESIDENTIAL BUILDINGS, WAYS TO DETERMINE THEM

Determining the minimum distance of wind turbines from residential buildings in most countries is done on the basis of simulations of environmental noise propagation. However, in some countries, despite the simulations carried out, there are impassable minimum values. In some countries these values are mandatory, in others they are negotiable at the level of local law. Therefore, the summary presented in Table 2.1 illustrates only the preferences of state governments regarding environmental protection assumptions around wind turbines. The table was made on the basis of a critical analysis of documents available on the internet resources.

Looking only at European conditions, one can see that minimum acceptable distances vary considerably, ranging from $10H$ to no minimum distance in national regulations. On the other hand, the minimum distance adopted in Hungary (12 km) is incomprehensible, which virtually excludes the possibility of investing in onshore wind energy in that country.

Such variation can also be seen outside Europe. In Australia, minimum distances vary from state to state, ranging from 1,000 to 2,000 m. For instance, the website of Australia's National Wind Farm Commissioner (https://www.nwfc.gov.au/observations-and-recommendations/governance-compliance) indicates that in the state of Victoria the minimum setback distance was not specified before 2011, after 2011 it was set as 2 km, then in 2015 it was changed to 1 km.

According to the American Wind Energy Association, minimum setback distances in the United States from residences, property lines, roads, environmentally or historically sensitive areas and other locations can be set by federal, state and/ or local authorities, depending on the specifics of the project. These can be a fixed

TABLE 2.1
Minimum Distances of Wind Turbines from Residential Buildings

Country	Minimum Distance	Comments
Belgium (Walloon region)	4H	H – Maximum height calculated from the ground
Denmark		to the tip of the blade in the highest position
Poland	10H	
Bavaria (Germany)		
Italy	6H or 500 m	
Estonia	500 m	–
France		
Greece		
Ireland		
Lithuania		
Netherlands	400–600 m	Depending on the region
Austria	750–1,200 m	Depending on the regional policy of the canton
Germany	1,000	Distance recommended by federal government, may be reduced by provisions of local state law
Romania		Distance recommended by state regulations, may be reduced based on noise simulations carried out
Belgium (Flanders)	–	No minimum distance in state regulations
Finland		
Norway		
Sweden		
United Kingdom		

distance or a distance relative to the height of the turbine (e.g., 550 m in Iowa; none in Vermont; 1.1–1.5HH in California).

In Canada, also the minimum distance from residential buildings depends on the state government (e.g., 550 m in Ontario, 1,000 m in Nova Scotia, 700 m in Saskatchewan).

Many examples as well as highly varied distances between wind turbines, and buildings in the United States and Canada can be found at https://ontario-wind-resistance.org/setbacks/.

As can be seen, in many countries the minimum distances that are taken into account are simply recommendations and have not been introduced by common law. In many cases, the authority to make decisions on the location of wind farms lies with local governments and local communities. In addition, the applicable guidelines are not rigid in nature and should be applied flexibly, examining each case individually.

2.1.3 SIMULATIONS OF AUDIBLE NOISE EMISSIONS

One of the environmental impacts of wind turbines analyzed in the report as part of the procedure for obtaining a decision on the environmental conditions of the project is noise emissions. This is the design stage of the wind farm. Accordingly, the analysis is based on simulation of sound wave propagation. One of the calculation models commonly used for this purpose in the world is the model described in the PN-ISO 9613-2:2002 standard. The method enables to predict noise under conditions that favor propagation (downwind). The accuracy of the method has been repeatedly tested by various researchers (Evans & Cooper 2012; Probst et al. 2013; Zagubień & Ingielewicz 2017) and has been determined to be $\pm 3\,$dB. When determining the range of noise emissions, two parameters are of key importance, i.e., the sound power level of the turbine and the effect of the ground type on the propagation of the sound wave. The sound power level is determined according to IEC 61400-11 with an accuracy of $\pm 2\,$dB (IEC 61400-11 2012). The influence of the ground type is determined in accordance with PN-ISO 9613-2:2002 by the A_{gr} (PN-ISO-9613 2002). The standard allows using two scenarios for determining the A_{gr}, the general method and the alternative method. In the general method, the A_{gr} is determined from formulas that take into account the ground index G. In the alternative method, the A_{gr} is determined by taking into account the distance between the source and receiver and the average height of the propagation path over the ground.

As indicated in Section 1.1, the acoustic power of wind turbines, given as the sum of aerodynamic and mechanical noise, always depends on the wind speed due to the nature of the device's operation. Modern wind turbines have the ability to adjust the nominal acoustic power of the turbine and use silencing serrations on the rotor blades. With a view to predicting noise emissions safely for human health, a restriction on the use of noise-reducing trailing edge serrations ("mod" settings) should be introduced at the design stage of the farm. The restriction would consist in leaving the possibility of applying a buffer of 3 dB. The designer should demonstrate in the acoustic analysis that there is, after the construction of the wind farm, the possibility of reducing the acoustic power of the turbine or using a technical solution on the turbine, reducing noise emissions by 3 dB.

It seems necessary to determine a uniform method for determining the value of the A_{gr} index adopted in simulations of wind turbine noise propagation. Two scenarios are possible: conducting calculations using only the alternative method (the ground index G is not determined) or conducting calculations using the general method and indicating how to determine the ground index G. The standard clearly indicates only two values for the soil index G: $G = 0$ for hard surfaces (concrete, paving, water, ice, etc.) and $G = 1$ for porous surfaces (all soil surfaces suitable for vegetation growth). In other situations, the percentage of porous soil is used. A more detailed description of the ground index G is provided in the EU Commission Directive 2015/996 (Directive 2015/996 EC). According to the directive (Directive 2015/996 EC) the value of the G index should be taken in line with Table 2.2.

The novel proposal for determining the ground index G was proposed in the article "Tłumienie gruntu w analizach akustycznych farm wiatrowych" ("Ground attenuation in acoustic analyses of wind farms") (Ingielewicz & Zagubień 2014). In this proposal, the determination of the value of the ground attenuation coefficient requires a thorough identification of the terrain, existing infrastructure, the presence of permanent and periodic water bodies and land use. For wind farm sites, situation and elevation maps, satellite maps, obtained information from local inspections and the Local Land Use Plan are analyzed. As a result of the analysis, the percentage of paved areas, flat areas covered with vegetation, areas of standing and flowing waters,

TABLE 2.2
Values of G Index for Different Ground Types

Ground Description	Type	G
Very soft (snowy or similar to moss-covered)	A	1
Soft forest cover (short, compacted, similar to heathland or overgrown with dense moss)	B	1
Non-compact, loose ground (peat, grass, loose soil)	C	1
Normal, non-compacted ground (forest cover, pasture)	D	1
Compacted field and gravel (compact lawns, park areas)	E	0.7
Dense, compacted substrate (gravel road, parking lot)	F	0.3
Paved surfaces (most normal asphalt types, concrete)	G	0
Very hard and compacted surfaces (compacted asphalt, concrete, water)	H	0

as well as the depression of the land where rainwater and snowmelt can accumulate are determined. It is assumed that standing and flowing water as well as rainwater and snowmelt can freeze during the winter months. Both the melting of the deposited snow layer and freezing in the winter season are slow and gradual processes (mild temperature drops or rises). The resulting water is mostly absorbed by the soil, and in part may collect and linger temporarily in existing land depressions and ditches. Thus, as a result of refreezing, the ice sheet will occur only on a limited area representing only a small percentage of the total area. The size of this area depends on the terrain. The authors proposed to calculate for the entire area of influence of the wind farm the percentages of hard ($G=0$) and porous ($G=1$) surfaces. Hard surfaces included: paved roads and squares, areas of standing and flowing water, land depressions where water can periodically accumulate. Other areas were classified as porous surfaces. From the analyses carried out in the abovementioned manner, the ground index ranges from 0.9 to 1.0.

As far as reducing the impact of noise on the people living in the vicinity of wind farms is concerned, it is proposed to use a uniform ground index of $G=0.7$ for simulations conducted at the stage of obtaining a decision on the environmental conditions of the project. This value is in line with recommendations used, for example, in Canada (Ministry of the Environment 2008). At the stage of performing environmental forecasts during the procedure of enacting the Local Spatial Development Plan, exclusively for the areas of the future wind farm, the ground index G should be omitted, and calculations should be carried out by the alternative method in accordance with the PN-ISO 9613-2:2002 standard (PN-ISO-9613 2002) and use the single-number value of acoustic power. At the stage of spatial planning, it is not possible to strictly determine the type of equipment installed in the future. If a local zoning plan is passed for an area that includes mixed-function areas, the outline of the wind farm should be set back 500 m from residential areas. Detailed proposals are discussed in Chapter 4.

The effect of the cumulative impact of several wind farms (i.e., investments in the immediate vicinity of the planned project) should be considered only for projects that are under construction or completed. An exception may be made for investments that are feasible on the basis of provisions in the Local Spatial Development Plan.

It is recommended to create a register of wind farm sites. The register should include installations under construction (in the construction phase) and completed, as well as areas provided for wind farms in the Local Spatial Development Plan.

2.1.4 POST-IMPLEMENTATION AUDIBLE NOISE ANALYSES

The purpose of post-implementation noise analyses is to assess the operating wind farm in terms of environmental and human risk based on in situ measurements. In addition, field measurements are a verification of the results of numerical calculations made at the design stage of the farm. Due to the specificity of acoustic phenomena occurring during the operation of wind turbines, it is advisable to supplement or develop a methodology for carrying out noise measurements of wind farms. The methodology for carrying out environmental noise measurements in Poland is set out in the Regulation (Journal of Laws 2021 item 1710, Appendix 7) (Regulation – emissions 2021). The recommended reference measurement methodology establishes a number of restrictions related to the atmospheric conditions prevailing during measurements, the location of measurement points, measurement sets, and defines the method of performing measurements.

The following changes, recommendations and additions to the reference methodology for performing noise measurements for wind turbines are proposed:

- Introduction of a recommendation to conduct measurements only by sampling method due to the variability of wind speed and other meteorological parameters throughout the day.
- The measurement time of the noise sample (immission – turbine noise plus acoustic background) enabling to eliminate the impact of interference and taking into account the variability of wind speed, set at 300 seconds, and reduced to 60 seconds if frequent interference occurs.
- The measurement time of the acoustic background sample, allowing the elimination of the influence of interference, set at 60 seconds, and reduced to 10 seconds if frequent interference occurs.
- Allowing multi-day monitoring measurements for evaluation and determining the results based on a selection of measurement samples from a wider period and statistical analysis of noise levels, correlated with measurements of non-acoustic parameters, which would help to reassure the public by providing noise evaluations for a wide variety of operating states and acoustic wave propagation conditions (a detailed description of this point of the methodology requires additional field studies, these methods are commonly used in the evaluation of aircraft noise and variability of noise emissions from industrial sources).
- Acoustic background measurements should be made only after stopping the turbines at the same control points where noise was measured during their operation. Measurement of background noise in the acoustic shadow of the residential building under study is acceptable, provided that there is only one turbine next to the building under study within a 1,000-m radius (then the influence of other turbines will be negligible during the background measurement).

- The location of control measurement points should be established on the borders of the parcels of the nearest acoustically protected areas, at a height of 1.5 m above ground level. It is necessary to exclude the possibility of measurement at the façade of the protected building in order to subtract a correction (−3 dB) related to sound reflections from the façade. The average wind speed during measurements at a height of 3.5 ± 0.5 m must not exceed 5 m/s.
- Prior to the measurements, the measurement team should read the acoustic part of the environmental impact report and the technical documentation of the wind turbines and determine at what wind speeds the given type of turbine reaches the maximum sound power level. If the acoustic data is related to the wind speed at a height of 10 m above ground level, it should be extrapolated to the height of the wind turbine nacelle.
- The noise measurement must be carried out under conditions of at least two different wind speeds measured at the nacelle, with measurements covering the average wind speed, from those recorded at all wind farm turbines, with a value of 70% of the speed causing maximum acoustic power or higher.
- A recommendation to conduct measurements in four seasons should be introduced, which will allow taking into account atmospheric conditions that vary throughout the year, such as humidity, temperature, wind pressure and direction, as well as variable ground cover affecting sound wave propagation. The measurements should be repeated in the following year, and if it is shown that permissible levels are not exceeded, it should be considered that the wind farm does not require further noise studies in its surroundings.
- Wind directionality should be taken into account. Measurements should be conducted at all control points in each season. The propagation of noise with the wind to the point should be taken into account, and enough measurement sessions should be carried out ensuring that at least once at a given measurement point the wind direction is within the $\pm 45°$ angle dilation, looking from the nearest wind turbine to the control point.
- There should be an obligation for the wind farm operator to make the information about the average wind speeds on individual turbines recorded during the performance of noise monitoring available to the measurement team.
- On the basis of the wind farm operator's computer records, it should be determined at what wind speeds were noise emission measurements at the height of the axis of individual turbines made. The range of averaged wind speed values, corresponding to the time when the measurements were made, should be given in the test report. It should be stated what turbine settings (mods) were used during the measurements.
- Nighttime is preferred during the measurements (lower permissible level, lower acoustic background, more stable atmospheric conditions).

    In the cases where the measured noise emission level is indistinguishable from the acoustic background, and at the same time the emission level does not exceed the values of the permissible levels at a given point, a note that in such a case the noise emission level at a given point does not pose a threat to the environment and people should be provided.

2.1.5 Infrasound and Low-Frequency Noise

In addition to audible noise, wind turbines are a source of low-frequency noise (LFN) emissions in the range of 20–200 Hz (Schmidt & Klokker 2014). Since many community complaints relate to LFN emitted by wind turbines, it is important to assess the impact of LFN on the health of residents living near wind farms. A study by Schmidt and Klokker (2014) found that there is no statistically significant evidence showing an association between wind turbine noise exposure and tinnitus, hearing loss, dizziness or headache.

According to Zajamšek et al. (2016) the main infrasound components of wind turbine noise can be measured at distances of many kilometers from the wind farm. However, it is usually at levels well below the normal hearing threshold.

A study by Onakpoy et al. (2015) found that wind turbine noise affects sleep and quality of life, but this relationship is unclear. It was shown that there is some evidence that the exposure to wind turbine noise is associated with an increased likelihood of nuisance and sleep problems. However, individual attitudes may influence the type of response to wind turbine noise. Thus, experimental and observational studies on the relationship between wind turbine noise and health status are justified. Important in the aspect in question are the results of a Finnish research project, which clearly shows that inaudible low-frequency sounds emitted by wind turbines are not harmful to human health, despite widespread concerns that they cause unpleasant symptoms (Maijala et al. 2020). The project involved surveys, sound measurements and provocative experiments. Some people reported various symptoms that they intuitively associated with infrasound emitted by wind turbines. In the survey, symptoms intuitively associated with wind turbine infrasound were relatively common within 2.5 km of the nearest wind turbine, and the spectrum of symptoms was wide. Many of the respondents who experienced symptoms also associated them with vibrations or the electromagnetic field of wind turbines. In the measurements, infrasound levels were similar to the levels typically found in urban environments. Prepared sound samples with the highest infrasound levels and amplitude modulation values were used in double-blind provocation experiments. The participants who had previously reported symptoms associated with wind turbine infrasound were unable to sense infrasound in the noise samples and did not find the infrasound samples more annoying than those with no infrasound components associated with wind turbines. The exposure to infrasound caused by wind turbines did not elicit physiological responses in any group of participants. It is clear from the study that there is a lack of scientific evidence of a potential link or studies focusing directly on the health effects of wind turbine infrasound.

Similar results were reported by Poulsen et al. (2018), who pointed out unequivocally that no convincing evidence was found for a link between the acoustic impact of wind turbines and heart attack or stroke in humans. The study found that noise from wind turbines did not affect sleep disorders and psychological stress (Yoon et al. 2016). In contrast, another study found that simulated infrasound had no statistically significant effect on the symptoms reported by people (Tonin et al. 2016).

The noise generated by wind turbines can resemble the sound of a long train passing by (Basner et al. 2011). It is created when a rotor blade passes the turbine tower, while the frequency depends on the frequency of passing of these blades (Pawlas et al. 2012). This periodic pulsing of noise in the audible band is erroneously assessed as infrasound (Leventhall 2011). According to Pawlas et al. (2012) the people exposed to noise from wind turbines may judge these sounds to be more bothersome than noise from other sources with the same parameters, such as road or industrial noise. An interesting study was conducted by Leventhall (2009), who showed that there are a number of non-acoustic problems that can lead to the perception of noise, but are not caused by it. In an earlier publication of his, Leventhall (2011) pointed out unequivocally that infrasound emitted by wind turbines is below the threshold of hearing and has no consequences for human health.

The opinion contradictory to the evidence presented above, based on a review of the world literature, is presented in the text by Barbara Lebiedowska titled "Farma wiatrowa jako sąsiad społeczności wiejskiej. Oddziaływanie akustyczne farm wiat-rowych" ("Wind farm as a neighbor to a rural community. Acoustic impact of wind farms"). The author states that there are no items in the literature that conclusively demonstrate the impact of wind turbines in terms of infrasound and low-frequency noise emissions on human health. Contrary to the above, there is a large body of literature available that indicates that noise with a frequency of less than 60 Hz generated by wind farms is impossible to be perceived by the human body (Maijala et al. 2020; Van Kamp & Van den Berg 2021), since this noise is below the level of human auditory and vibrational perception.

As already indicated in earlier sections of this study, there are both natural and anthropogenic sources of noise, emitting infrasound, in the human environment. In this regard, wind turbines are not isolated, and from this point of view, wind turbines are not exceptional devices. However, it should be pointed out that every human being is exposed to infrasound occurring in their environment, regardless of where they are (Zagubień & Wolniewicz 2016, 2017). It is true that after exceeding certain intensity levels, infrasound can potentially contribute to nuisances amounting to excessive fatigue, discomfort or other physiological functions. However, all of these phenomena, as also highlighted earlier, are perceived and described in a subjective manner and depend on individual sensitivity.

It has been found that only exposure to very high levels of infrasound noise can be hazardous to health (Helbin 2008). Thus, when assessing the exposure to infrasound noise, including that from the operation of wind turbines, it is necessary to know the sound pressure levels of infrasound noise in the environment and attempt to assess the risk on this basis. Chapter 1 showed that infrasound noise recorded around wind farms is at levels close to the emissions of natural sources and below the threshold of human perception.

Currently, Poland lacks the regulations governing the assessment of infrasound and low-frequency noise (LFN) in the environment. The problem does not apply only to the impact of wind turbines. In the occupational, municipal and outdoor environments also there is a lack of any regulation in this regard. There is an urgent need to regulate how infrasound and LFN assessments are carried out in Poland.

Denmark has the clearest regulations on infrasound noise assessment (Regulation Denmark 2011, 2019).

Although currently there is no clear evidence of the impact of infrasound from wind turbines on human health, the introduction of solutions modeled on the Danish ones, after appropriate clarification and taking into account Polish conditions, seems justified. This topic requires in-depth research, within the framework of which recommendations can be formulated for a reference methodology for performing calculations and measurements of low-frequency and infrasound noise in relation to wind turbines. A simplified assessment procedure example could be as follows:

a. the source emissions should be determined on the basis of the noise spectrum specified by the manufacturer. In the absence of relevant data, measurement of the noise spectrum in the immediate vicinity of the turbine should be conducted using, for example, a simplified procedure for determining the sound power level of the source based on IEC 61400-112012,

b. the typical standardized sound insulation of the building envelope should be adopted in the calculations (in Denmark this value is defined on the basis of a series of tests; a Polish study would be needed to define this value for typical residential development),

c. the noise level inside the residential premises should be assessed on the basis of calculations, based on the data indicated in pp.a and pp.b and the legally defined formulas (this approach in practice allows separating the result of the assessment from local influences, such as, for example, domestic noise),

d. at the stage of control measurements, modifying the methodology used in Denmark, the measurement of low-frequency noise can be carried out at a distance of 3 m from the façade of a residential building and the determination of power can be omitted (pp. a),

e. permissible levels for low-frequency and infrasound noise should be established (in Denmark, at night, in residential areas, it is 20 dB(A) and 85 dB(G)). In Poland, due to other applicable standards, there is a need to introduce appropriate limits.

2.2 SHADOW FLICKER

Shadow flicker should be determined as part of pre-construction activities for a wind farm. Reports can be provided to facilitate understanding the possible impact of shadow on properties, buildings and roads.

The intensity of the shadow flicker effect, and thus its perception by humans, depends on a number of factors, which include, first and foremost (Ove Arup and Partners 2004):

- tower height and rotor diameter;
- distance of the observer from the wind farm;
- time of year;
- cloud cover;
- presence of natural barriers (e.g., trees) between the wind turbine and the observer;

- orientation of windows in buildings located in the shadow flicker zone, relative to the wind turbine generating the phenomenon;
- lighting in the room.

Determination of the degree of annoyance of the optical impact in question is possible with the use of specialized computer software (e.g., WindPRO, WindFarmer, WindPass), which has a number of options for determining what the shadow incidence will be for specific environmental conditions. The calculations show in a precise way how often and at what intervals a given recipient will be affected by the shadow generated by the wind farm (Nawrotek 2012).

The computer programs indicated above carry out a simulation by analyzing the position of the sun relative to the wind turbine, at minute intervals for a period of the whole year. In order to carry out the calculations, it is necessary to enter the following input data into the program: the technical parameters of the wind turbine, the location of residential buildings, or the location and orientation of windows in these buildings. The calculation model used is also based on meteorological data. Using specialized computer programs, it is possible to carry out two types of calculations – actual shadow incidence and "worst case." The first situation is based on actual data from a meteorological station. "Worst possible case" is a situation that assumes any physical phenomena that aggravate shadow fall, such as lack of cloud cover. The result for the "worst possible case" determines the maximum duration of the so-called "shadow flicker effect" that a particular wind farm can produce. Having obtained the results from such calculations, one can be sure that the shadow impact will not be greater for a particular project (Nawrotek 2012) (Figure 2.1).

Shadow flicker

over 40

30 - 40

20 - 30

10 - 20

5 - 10 (h/year)

FIGURE 2.1 Butterfly map showing the effect of shadow flicker resulting from the operation of a wind turbine (Nawrotek 2012).

The results from the calculations can be further presented in the form of a so-called butterfly map. The butterfly map shows the sum of minutes for which each raster cell on the map will be shaded during the year.

The user can interactively define the resolution of the raster cells and the time steps for which the butterfly map is calculated. The number of minutes for which each raster cell is shaded is displayed in a window when a raster cell on the butterfly map is clicked (M.O.S.S. Computer Grafik Systeme GmbH 2015). An example of a projection from a specialized software suite is shown in Figure 2.2. In contrast, the earlier part of this publication presented graphical results obtained using the WindFarmer software.

2.3   ELECTROMAGNETIC FIELD

The study of the effects of slowly varying electromagnetic fields on human health has been one of the most widely conducted epidemiological studies in history (Szuba 2008). This was due to the prevalence of electrical networks and installations with a frequency of 50 Hz (60 Hz) and also from the belief that their effects are harmful, particularly with regard to certain types of cancer. Although the results of these

FIGURE 2.2   Simulation of the so-called shadow flicker effect in the WindPASS software suite (M.O.S.S. Computer Grafik Systeme GmbH 2015).

studies are inconclusive, it is certain that long-term (many years) exposure to high levels of 50 Hz magnetic field strength can be a potential carcinogenic factor. Thus, on the one hand, 50 Hz magnetic field is classified as a factor "with probable carcinogenic effect"; on the other hand, a critical review of the available scientific literature provides arguments both for the existence of associations between 50/60 Hz magnetic field and cancer risk, and arguments against such a concept (Szuba 2008). However, it should be noted that the regulations in force in Poland – the Regulation of the Minister of Health of December 17, 2019 (Regulation – electromagnetic fields 2019) provide a sufficient level of safety for the people under the influence of such fields, especially since the requirements for Poland are more restrictive in this regard than in other countries. This is confirmed by Table 2.3.

A synthetic compilation of opinions on the effects of permanent and slow-variable magnetic fields on living organisms is presented in Table 2.4.

According to the paper (McCallum et al. 2014), magnetic field measurements were carried out in the vicinity of fifteen 1.8 MW wind turbines, two substations, various aboveground and underground collectors, as well as transmission lines and nearby houses, under various operational scenarios (high wind, low wind and windless conditions). They found that magnetic field levels decreased rapidly with increasing distance, and that none of the potential field sources affected magnetic field levels in nearby homes. In fact, the authors note that magnetic field levels near the wind

TABLE 2.3
List of Regulations on Permissible Values of 50 Hz Magnetic Field

Country or International Organization	Permissible Value of Electromagnetic Field Strength H (A/m)	Comments and Recommendations on the Application of the Permissible Value
Austria	80	So-called Reference Level
Belgium	–	The permissible value applies only to the electric component of the field (E)
Croatia	80	Value identical to that recommended by ICNIRP
Czech Republic	80	Value identical to that recommended by ICNIRP
Denmark	–	No regulations; recommended so-called precautionary approach, i.e., limiting the construction of overhead lines close to residential buildings and, accordingly, the erection of residential buildings in the vicinity of existing lines
Estonia	80	Recommended value for long-term exposure according to Recommendation 1991/519/EC
Finland	80	Recommended value for long-term exposure
	400	Recommended value for short-term exposure
France	80	Recommended value for exposure in fields generated by new and reconstructed facilities and installations of the power system operating under normal conditions
Spain	–	No regulations for 50 Hz fields

(Continued)

TABLE 2.3 (*Continued*)
List of Regulations on Permissible Values of 50 Hz Magnetic Field

Country or International Organization	Permissible Value of Electromagnetic Field Strength *H* (A/m)	Comments and Recommendations on the Application of the Permissible Value
Netherlands	96	Recommendation of the Health Council of the Netherlands (controlled value)
	80	Government recommendation for local authorities and power companies
	0.32	Value averaged over 1 year; government recommendation for local authorities and power companies for places inhabited by children over long periods of time (apartments, schools); recommendations for new overhead lines and new buildings located in the vicinity of overhead lines, provided this is feasible
Germany	80	Recommended value for exposure in fields generated by facilities and installations of the power system (overhead and cable lines, substations, etc.)
	160	As above, but with exposures of less than 1.2 hours per day
Poland	60	Staying in the field without time limitation
Portugal	80	Recommended value for long-term exposure
Slovenia	8	In areas of special protection (locations of residential buildings, schools, hospitals, recreation centers, etc.)
	80	In other areas
Switzerland	0.8	In the case of construction of new overhead lines, the level recommended in areas of special protection (locations of residential buildings, schools, hospitals, recreation centers, etc.) provided that this does not entail excessive costs for the implementation of the project, the so-called "precautionary approach"
	80	In other areas
Sweden	80	Recommendations of the Swedish radiological protection service
	–	For new lines, the so-called "precautionary approach" is recommended, aiming to maintain natural magnetic field levels, provided that this does not entail excessive costs for the implementation of the project
Hungary	80	Determined at a height of 1.5 m agl
United Kingdom	80	So-called reference level
Italy	80	So-called exposure limit
	8	So-called attention level; daily average when staying in the field generated by overhead lines for at least 4 hours a day
	2.4	So-called qualitative target; daily average when staying in the field produced by newly constructed overhead lines
ICNIRP	80	Staying in the field without time limits (International Commission on Non-Ionizing Radiation Protection)
Recommendation for European Union countries	80	Recommended value for long-term exposure according to Recommendation 1991/519/EC (Council of European Union)

TABLE 2.4

Effects of Permanent and Low-Frequency Magnetic Fields on Living Organisms

Field Intensity (A/m)	Permanent Magnetic Field		Variable Magnetic Field with a Frequency of 50 Hz	
	Effects on Living Organisms	Need for Human Protection		Effects on Living Organisms
Under 0.8	No effects at all; magnetic fields of this intensity are not perceptible by living organisms	No indication for restriction and human protection	Under 0.8	No effects at all; magnetic fields of this intensity are not perceptible by living organisms
0.8–8.0	No effects at all; magnetic fields of this intensity are not perceptible by living organisms	No indication for restriction and human protection	0.8–8.0	No effects at all; magnetic fields of this intensity are not perceptible by living organisms
8.0–80	No physiological effects; this is the range of the earth's natural magnetic field; perceived by some animals	No indication for restriction and human protection	8.0–80	No physiological effects; this is the range of the earth's natural magnetic field; perceived by some animals
80–800 (0.1–1 mT)	No physiological effects	No indication for restriction and human protection	80–800 (0.1–1 mT)	No physiological effects
800–8,000 (0.1–10 mT)	No physiological effects	No indication for restriction and human protection	800–8,000 (0.1–10 mT)	No physiological effects
8000–80,000 (10–100 mT)	No effect on the human body (employee studies)	Restrictions for the population; WHO (1984); and Europe (1995); entire day – 32 kA/m (40 mT)	8000–80,000 (10–100 mT)	No effect on the human body (employee studies)
80,000–8 m (1–10 T)	Possibility of physiological changes in some organs under the influence of fields stronger than 1.6 MMA/m (2T)	Restrictions for employees: 160–1,600 kA/m	80,000–8 m (1–10 T)	Possibility of physiological changes in some organs under the influence of fields stronger than 1.6 MMA/m (2T)

turbines were lower than those produced by many typical household electrical appliances and well below existing regulatory guidelines for human health.

The above observations confirm the results of a study (Aris et al. 2020) conducted in Greece – the authors found that the electromagnetic field around wind turbines was similar or lower compared to that found in urban areas and well below national and international safety limits.

Similar conclusions can be drawn from measurements made by the authors of this monograph around transformer/switching stations and turbine towers at selected wind farms in Poland. The results of these studies are summarized in Tables 2.5 and 2.6. The measurement methodology was adopted in accordance with the Regulation of the Minister of Climate (Regulation – electromagnetic fields 2020). Measurements of the electric component of the electromagnetic field were made at each measurement point at a height of 2.0m above the ground surface, in three mutually perpendicular directions x, y, z, using a boom for the electric component meter to eliminate interference. Measurements of the magnetic component of the electromagnetic field were made at each measuring point in measuring risers at heights of 0.3–2.0m above the ground surface, in three mutually perpendicular directions x, y, z (Tables 2.5 and 2.6).

TABLE 2.5
Electromagnetic Field Levels at Wind Turbine Towers

Direction of Measurement	Measurement Results		Permissible Level	
	Value (V/m)	Measurement Height (m)	Residential Area (V/m)	Undeveloped Area (V/m)
Electrical Component E				
X	2.8–4.3	2.0	1,000	10,000
Y	2.9–3.9			
Z	3.2–4.7			
Magnetic Component H				
X	0.03–0.05	0.3–2.0	60	60
Y	0.03–0.05			
Z	0.03–0.05			

TABLE 2.6
Electromagnetic Field Levels at Transformer/Switching Station

Direction of Measurement	Measurement Results		Permissible Level	
	Value (V/m)	Measurement Height (m)	Residential Area (V/m)	Undeveloped Area (V/m)
Electrical Component E				
X	3.7–190.0	2.0	1,000	10,000
Y	2.9–220.6			
Z	13.2–500.6			
Magnetic Component H				
X	0.02–0.05	0.3–2.0	60	60
Y	0.03–0.28			
Z	0.03–0.06			

None of the wind farms surveyed recorded electromagnetic field levels exceeding the residential permissible levels of 1 kV/m for the electric component and 60 A/m for the magnetic component.

On the basis of the considerations made and the results of measurements, it can be concluded that the impact of wind turbines on human health in terms of electromagnetic fields should be considered as impacts in the ELF (extra low frequencies, 50 Hz) type of fields, using dedicated standards.

Due to the height of wind turbine masts, impacts from generators and other equipment in the turbine nacelle may not be considered for people on the ground. The electric and magnetic components of the electromagnetic field that may affect humans are generated by electrical equipment that outputs power from the wind turbine and feeds it to the distribution substation (MV or 110/SN kV). Computational analyses and measurements (including those made by the authors of the monograph) show that the intensity values of these fields are at least several or more times lower than the regulatory limits (Regulation of the Minister of Health dated December 17, 2019). Surface compaction of cable lines or their proximity to buildings can result in higher values of magnetic field strength, but it is impossible to exceed the critical values if obvious design errors in the construction of the lines and the requirements of the standard (SEP-E 004 2022) are excluded. Transformers and distribution equipment include partitions and fences that prevent people from approaching close enough for the generated electromagnetic field to cause any health consequences (Electromagnetic Compatibility 2004; Krug & Lewke 2009).

Thus, if the internal network of the wind farm and its other power infrastructure are built in accordance with the engineering principles analogous to those for cable lines used in the distribution of electricity at the level of municipal and industrial consumers, then the electromagnetic field associated with the operation of the wind farm does not have a negative impact on human health, according to the current Polish regulations.

The impact of workers staying briefly in the nacelles of wind turbines to perform repairs and maintenance should be assessed separately. In these cases, the recommendations of the Central Institute for Occupational Safety and the operating instructions issued by the manufacturers of each type of equipment should be strictly followed.

2.4   VIBRATIONS AND OSCILLATIONS

The methodology for determining the degree of maximum intensity of negative impacts of wind turbines on human health and acceptable standards for vibration currently used in Poland provide an adequate level of safety (PN-B-02170 2016; PN-B-02171 2017).

According to the International Standards Organization (ISO 2631 1997), three thresholds of human sensitivity to vibration exposure can be distinguished: I – perception threshold, II –annoyance threshold, III – tolerance threshold.

From a study conducted by Nguyen et al. (2020) an assessment based on AS 2670-2 and BS 6472-1 (Australian standards are identical to ISO 2631) showed that the vibration levels on the floor of dwellings are unlikely to cause discomfort or negative comments (PN-B-02171 2017). The vibration levels of wind farms measured inside residential buildings located nearby showed that the values obtained were below acceptable levels. Moreover, a comparison of vibration signals with the basic acceleration curve specified in AS 2670-2 showed that the measured vibration levels at the bed frame and floor were too low to cause discomfort.

As indicated by Gaj and Blaszczak (2013) the most harmful to human health are the vibrations of very low frequencies (from a few to tens of Hz). They conducted a study in which the closest household was at a distance of 300 m from a wind turbine. On the basis of the results, Gaj and Blaszczak concluded that in the range of very low frequencies (and therefore particularly harmful to humans), the measured vibration values were small. They also indicated that it is difficult to find negative effects of wind turbines on human and animal health (Figure 2.3).

Household 1 was located at a distance of 300 from the wind turbine, while, for example, household 5 was located at a distance of 900 m from the turbine. It can be seen from the graph that the overall vibration levels obtained are low, while there is no significant difference between the values obtained for individual houses. Importantly, none of the distances studied exceeded the threshold values for mechanical vibration (threshold value of 2.5 m/s^2) and general vibration (threshold value of 0.5 m/s^2), as defined for workers, i.e., those most exposed to the negative impact of vibration from a wind turbine, as referred to in the regulation of the Minister of Economy and Labor (Regulation – noise 2005).

In recent years, the operation of wind turbines has also been linked to causing various symptoms and diseases, among which is vibroacoustic disease (VAD). Alves-Pereira and Castelo Branco (2007) issued a press release suggesting that living near wind turbines leads to the development of vibroacoustic disease in residents of nearby homes. However, the research was not published in a peer-reviewed journal or subjected to a detailed scientific review. They were only presented at a conference. Moreover, given where they were disseminated and the limited availability of data, it is difficult to assess whether the information provided by Alves-Pereira and Castelo Branco is reliable and valid. This is because these studies do not include noise measurements, but only measurements of the distance from the study participants to the nearest turbines. Thus, the sample adopted for the study does not contain an adequate

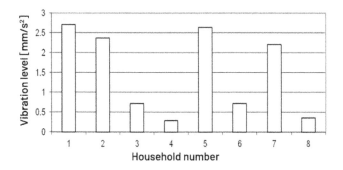

FIGURE 2.3   Comparison of vibration levels for selected households (Gaj & Blaszczak 2013).

statistical representation of potential health effects. It should be emphasized that unlike the questionnaires used by Pedersen et al. (2004, 2007, 2008), the purpose of the study was not hidden from the participants.

However, it should not be overlooked that of the 35 published papers on vibro-acoustic disease, a minimum of one of the above two authors is a co-author in 34. As with the work of Pierpont (2009) and the wind turbine syndrome she describes, authors Alves-Pereira and Castelo Branco are accused of failing to make the connection between vibroacoustic disease and exposure to wind turbines, and of being biased in their selection of subjects for the study.

Chapman and St. George (2013) point out that vibroacoustic disease has received virtually no scientific recognition outside the group that coined the term. Claims available to the public that wind turbines cause the aforementioned disease should be considered unsubstantiated. This state of affairs may contribute to the nocebo effect among those living near the turbines. This effect, in turn, includes all those circumstances in which there is a deterioration of human functioning under the influence of the use of a therapeutically inert agent or other therapeutic measures, but also the interactions undertaken between the patient and staff (Colloca 2017).

Although it is generally accepted that ground vibrations caused by wind turbines have no detectable effect on humans, further research work is needed to determine the effect of wind farms on ground vibrations and the distance at which they should be located from sensitive seismic measuring stations that warn of earthquakes or volcanic eruptions (Marcillo & Carmichael 2018).

2.5   MECHANICAL IMPACTS – ICE PIECES, BLADE FRAGMENTS

Certainly, hitting a person with a mechanical element or a piece of ice is dangerous to their life. It is estimated that a fatal danger arises when hit by an ice piece with an energy of more than 40 J, which corresponds to a piece with a mass of more than 0.2 kg, falling from a height of 30–50 m (Bresden et al. 2017).

Thus, being in the vicinity of wind turbines (even when they are stopped) involves danger from the risk of ice and snow clumps detaching from the blades. Estimating the level of this danger is the subject of a number of scientific studies and analyses, as well as advanced monitoring by farm operators and turbine manufacturers. One of its measures is the Localized Individual Risk per Annum (LIRA) (Bresden et al. 2017; Robinson et al. 2013). It determines the probability of a fatal accident to a person in a given zone per year.

It is estimated (Bresden et al. 2017) that the risk of being dangerously hit by a piece of ice for a person outside a circle with a diameter of $2H$ (the height of a wind turbine tower) is less than 10^{-6} (per year). A comparison of this value with the risk from other technical infrastructure objects and various human activities is given later in the book.

The potential impact on human health caused by ice throw from rotor blades was also noted in the work of Knopper and Ollson (2011) and Knopper et al. (2014).

Taking the problem of the risk of a person being hit by ice pieces from a wind turbine very seriously, it should be noted that a report (CWIF 2020) compiling wind turbine accidents from 1980 to 2020 shows that such accidents constituted 2.7% of the total number of accidents.

2.6 CATASTROPHIC FAILURES AND FIRES

A catastrophic failure of a wind turbine caused by damage to one of the blades and then continuing in the form of gradual destruction of subsequent blades and culminating in the collapse of the entire tower poses a very serious threat to the lives of those in the vicinity. Hazard assessment is made by analyzing risk indicators, which are then compared with risk indicators determined for other areas of human habitation and other types of human activity (Rausand 2011). The most common indicator is the annual risk of fatality.

In comprehensive analyses of risks and hazards from wind farms, the most significant element to be assessed is the fracture of a blade or the detachment of its significant part (50%, 33%, 25% of its length) (Robinson et al. 2013). The incidence of a major wind turbine blade failure (complete or partial break) is estimated at 10^{-3} to 10^{-4} per year. This means that for 3,000 turbines (9,000 blades) per year, about 10 blades can be seriously damaged. Their fall trajectory and the probability of hitting directly (a person) or indirectly (a residential or commercial building) is a complex probabilistic problem. The use of Monte Carlo analysis enables to determine indicators referred to as conditional LSIR (Location Specific Individual Risk), with the term "conditional" meaning that a blade break is assumed to have occurred. In other words, a fatal event resulting from a wind turbine failure is associated with the logical product of two events – a broken blade and a person near the turbine being hit by the broken piece. The result is a risk originating from wind turbines and involving a direct or indirect threat to the life of a person who is outside a circle with a radius of $2H$ (Robinson et al. 2013) (H – height of the turbine tower in meters, including the axis of the turbine). It should be considered in juxtaposition with other risk factors brought by modern civilization (Table 2.7).

TABLE 2.7

Estimation of the Annual Level of Risk of Death due to the Impact of a Blade (or Its Fragment) for a 2.3 MW Wind Turbine as Well as Other Human Events and Activities

Source of Risk Leading to Death	Annual Risk of Death of a Person	Assumptions
Wind turbine – direct impact of blade or its fragment (blade failure and impact combined)	10^{-9}	At a distance greater than $2H$ from the wind turbine tower
Wind turbine – indirect impact of blade or its fragment (blade failure and impact combined)	10^{-8}	At a distance greater than $2H$ from the wind turbine tower
Malignant neoplasm	2.6×10^{-3}	Averaged over population, England and Wales 1999
Lightning	5.3×10^{-8}	England and Wales 1995–1999
Mining industry	1.09×10^{-4}	United Kingdom 1996–2001

(Continued)

TABLE 2.7 (*Continued*)
Estimation of the Annual Level of Risk of Death due to the Impact of a Blade (or Its Fragment) for a 2.3 MW Wind Turbine as Well as Other Human Events and Activities

Source of Risk Leading to Death	Annual Risk of Death of a Person	Assumptions
Construction industry	5.9×10^{-5}	United Kingdom 1996–2001
Agriculture	5.8×10^{-5}	United Kingdom 1996–2001
Service sector	3.0×10^{-6}	United Kingdom 1996–2001
Amusement park	4.8×10^{-9}	Assuming four trips per year, United Kingdom 1996–2000
Road accidents (all forms)	6×10^{-5}	United Kingdom 1999
Rail travel accidents (per passenger trip)	2.3×10^{-8}	Fatality rate one per passenger trip, United Kingdom 1996–1997
Rail travel accidents (annual risk – commuting) Air accidents (per passenger trip)	10^{-5}	Annual risk of death: two trips per day, 45 weeks per year. Mortality rate per trip per passenger, United Kingdom 1991–2000
Air accidents (annual risk – holiday trips, per passenger)	1.60×10^{-8}	Annual risk of death: two flights per year
Work in industry	8×10^{-6}	Norway 2001
Work in mining	109×10^{-6}	Norway 2001
Work in construction	59×10^{-6}	Norway 2001
Work in the metal industry	29×10^{-6}	Norway 2001
Work in the electronics industry	2×10^{-6}	Norway 2001
Work in the service industry	3×10^{-6}	Norway 2001
Death from lightning strikes	8×10^{-7}	Norway 2001

For the sake of comparison, another type of civilization risk can be indicated, shown in COIB (2017). It reports the risk of a plane hitting a nuclear power plant when it is located 8 km from the airport. It amounts to 10^{-7}.

If adequate distance is maintained from the wind turbine tower, catastrophic phenomena associated with serious damage to the tower are reduced to a level, defined as the annual risk of a fatal event, of less than 10^{-7}. On the basis of Robinson et al. (2013), Barclay (2012), Uadiale et al. (2014) and Haugen (2011), it can be concluded that Seifert's formula, originally developed for determining the critical range of ice piece throw, provides a level of safety of this value. For the contemporary turbine models, this is a distance of about 300 m. Determining the exact risk distribution for the models selected in detail, the location of the towers, wind speed and direction distributions, requires collecting failure rate data and conducting Monte Carlo simulation studies. The results presented by Robinson et al. (2013) and shown below confirm that the risk at distances above 300 m is low.

Wind turbine fires in the past have inevitably led to the loss of entire facilities. Hence, the significant development of techniques to reduce their risk (active and passive) and the development of SCADA (Uadiale et al. 2014) monitoring system minimize the risks to humans by maintaining appropriate distances from the turbines and their towers.

Determining the exact risk distribution for detailed selected turbine models, tower locations, wind speed and direction distributions requires collecting data on failure rates and conducting simulation studies using the Monte Carlo method. The results cited below confirm that the risk at distances above 300 m is low (Robinson et al. 2013). They show the risk of a fatal impact (direct or indirect) on a human being located in a square ((1 m × 1 m) and (5 m × 5 m)) in the area of a 350 m × 350 m rectangle. It can be seen that outside the circle with a radius of 300 m, blue (i.e., risk amounting to 10^{-7}) is the dominant color (Figure 2.4).

The analysis of the results obtained allows concluding that the risk of fatal impact on a person as a consequence of a wind turbine failure is two or three orders of magnitude lower than the risk from other elements of technical infrastructure and the risk from their occupational activity. The basic prerequisite for obtaining such low risk values is to maintain an adequate distance from the wind turbine tower. In the publications cited above, this is defined as $2H$ (i.e., more than 200 m). However, even being in the vicinity of a wind turbine tower involves the risk rate comparable to daily train rides to work (Table 2.7).

Wind turbine fires in the past have inevitably led to the loss of entire facilities. Hence, the significant development of techniques to reduce their risk (active and passive) and the development of the SCADA monitoring system minimize the risks to humans by maintaining appropriate distances from the turbines and their towers. It should also be emphasized that manufacturers of contemporary turbines are widely introducing a system of remote monitoring of their operation (SCADA) and through a system of sensors (in a single 3–4 MW Vestas turbine there are over a thousand sensors (Vestas Wind Systems 2019; Performance Specification Vestas 2019)) they are able to remotely identify the condition of the equipment, long before serious disturbances occur. Procedures and measures to reduce the danger of fire hazards are adopted as standards for the construction and operation of wind turbines (CFRA Europe 2010).

Wind turbine fires in the past have inevitably led to the loss of entire facilities. Hence, the very significant development of techniques to reduce their risk (active and passive) and the development of SCADA (Uadiale et al. 2014) monitoring system minimize the risks to humans by maintaining appropriate distances from the turbines and their towers.

According to Uadiale et al. (2014) and WHO data (Palmer 2018) the probability of a wind turbine fire per year (according to reported cases) is 6×10^{-5}, although the authors believe that realistically it may be higher (the successfully extinguished fires are not reported). Thus, for 3,000 turbines, less than one per year is at risk of fire. A significant risk (at 300×10^{-6}) is connected with the work of the people who build

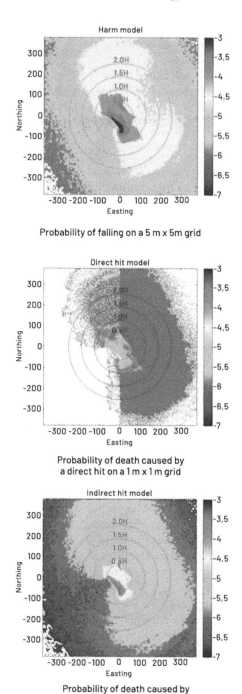

FIGURE 2.4 Simulation results of the risk of a fatal impact with a fragment of one-third of a blade (dark gray 10^{-7}, dark black 10^{-3}) (Robinson et al. 2013).

wind farms and are involved in their operation (work at height, mechanical injuries, effects of fires), as indicated by the authors of the report of the European Agency for Safety and Health at Work and the data found in CWIF (2020). The key to avoiding accidents is training, protective equipment, development of workers' competence, exchange of information and installation of appropriate alarm systems (EU-OSHA).

3 Wind Power Investments

3.1 LEGAL FRAMEWORK

In Poland, the dynamic development of wind power was recorded at the beginning of the 21st century. According to the Energy Regulatory Office (2020) installed capacity increased from nearly 85 MW in 2005 to just over 5,000 MW in 2015. This represents an increase of 4,900 MW in 10 years.

The presence of wind farms can contribute to conflicts of both environmental and social nature. The national network of protected areas, including the Natura 2000 network, as well as the dispersed development characteristic of rural areas, resulted in numerous challenges in the context of selecting the location for the implementation of wind investments. The selection of wind farm sites must be preceded by properly conducted forecasting of environmental and social impacts. Thus, wind facilities should be located based on one of the most important principles, namely taking into account the principle of sustainable development, as mentioned both in Article 5 of the Constitution of the Republic of Poland of April 2, 1997, as well as Article 3(50) of the Act of April 27, 2001 – Environmental Protection Law (2001).

The implementation of wind farms is related not only to the technical aspect of the investment but primarily to the process of acquiring investment decisions. This is a complex and lengthy process, sometimes taking even several years.

Despite such a significant development of wind energy in Poland, it is pointed out (Zajdler 2012) that the industry has faced numerous challenges from the very beginning, resulting from, among others:

- energy regulations;
- environmental regulations;
- planning regulations;
- tax regulations.

3.1.1 ENERGY

The basic legal act regulating the development of onshore wind energy in Poland is the Act of February 20, 2015, on Renewable Energy Sources (2015). The act implements into Polish Law Directive 2009/28/EC of the European Parliament and of the Council of April 23, 2009, on promoting the use of energy from renewable sources amending and subsequently repealing Directives 2001/77/EC and 2003/30/EC5 (Directive 2009/28 EC). In terms of onshore wind energy, the RES Act regulates, among other things, the mechanisms and instruments supporting electricity generation in RES installations, including the auction support system and the support system in the form of certificates of origin, or the rules for issuing guarantees of origin for the electricity generated in RES installations. While these guarantees do not give

DOI: 10.1201/9781003456711-4

rise to property rights, as is the case with certificates of origin, they do constitute a certification to the end customer that the volume of electricity injected into the grid indicated in this document was generated in RES installations.

According to Zajdler (2012) the main regulatory problem in the energy field relates to the rules of connection to the grid, the stability of the support system or access to information on the operation of the grid. According to Article 7, paragraph 8d3 of the Law of April 10, 1997 (Energy Law 1997) an energy company engaged in the transmission or distribution of electricity may refuse to issue the conditions for the connection of a renewable energy source installation in the absence of technical or economic conditions for connection. However, it should not be overlooked that despite repeated amendments to the above law over the course of several years, no definition of the lack of technical and economic conditions for connection of sources to the grid has been introduced. Thus, the regulations do not provide sufficient transparency in these procedures.

Although Zajdler (2012) described as unstable the system of financial support for energy produced from RES, it should be pointed out that the situation in the above regard has changed. In order to meet the EU targets and to ensure continued growth in installed capacity from renewable energy sources, two support mechanisms for wind power were introduced into the legal order in 2015, i.e., the green certificate system and the auction support system, which is to eventually replace the green certificate system.

The system of so-called green certificates has been implemented in Poland since October 1, 2005. Initially, it was regulated in the Energy Law, but after the RES Act came into force, the relevant legal norms in this regard can be found there. Importantly, after the adoption of the auction system (which will be discussed in more detail later) as the basic support model, the system of origin certificates is being gradually phased out. It is characterized by the fact that new installations cannot be covered by this support system and the gradual exit from the system of green certificates of installations after the end of the 15-year support period. It can be used by those RES installations where electricity was first generated before July 1, 2016.

The green certificate system is a quantitative support mechanism for the production of electricity from renewable energy sources. This means that energy from renewable sources receives appropriate certificates for every 1 MWh of electricity produced and fed into the grid. These certificates are issued by the president of the Energy Regulatory Authority at the request of the generator. Certificates of origin may be registered by the Polish Power Exchange. Property rights arising from certificates of origin then become the subject of trading on the commodity exchange, as well as can be sold in direct transactions. A significant difference between the auction system and the green certificate system is that the amount of support in the case of certificates is the same for all renewable energy producers. Moreover, the price is not known in advance, as the price of property rights fluctuates over time. Demand for green certificates, in turn, is guaranteed by the statutory obligation to purchase and present to the president of the Energy Regulatory Office for redemption a certain amount of green certificates. If this obligation is not fulfilled, the entity (generally an energy company) must pay a so-called replacement fee or administrative penalty. It is worth adding that installations that have so far been covered by the system of certificates of origin can switch to the auction system. If they win the auctions dedicated to existing installations, the assistance under the system of certificates of origin comes to an end (Polish Wind Energy Association 2022).

The auction system is currently the primary support mechanism for renewable energy source installations. As indicated above, it was intended to replace the system of certificates of origin, and this is due to the adoption of the principle that installations of renewable energy sources, in which the first generation of electricity is to take place or has taken place after July 1, 2016, can only benefit from the auction system. The introduction of the auction system into the Polish legal order was preceded by the European Commission's decision of December 13, 2017, which recognized the form of state aid in question as compatible with the internal market. The level of support is determined through an auction process, in which aid is granted in the form of a variable premium to the market price based on a differential contract for a specified period of support. The advantage of this form of support is its stability and predictability, which facilitates, among other things, the acquisition of external funding for the implementation of renewable energy installations, including wind farms. *Stability is associated with a predetermined period of support, which cannot be longer than 15 years from the date of the first sale of electricity after the closing date of the auction session (until now, the period has always been 15 years). Predictability, on the other hand, is associated, first, with the energy sale price, which is valid for the entire support period and is derived from the bid winning the auction and is then indexed annually for inflation, and second, with the total amount of electricity to be sold under the auction system, which is also stated in the bid* (Polish Wind Energy Association 2022).

A problem in terms of energy regulation for the implementation of RES investments, including wind farms, is that at the stage of issuing the certificate of admission to the auction, the president of the Energy Regulatory Office verifies whether the installation has valid connection conditions or has a grid connection agreement. In view of the numerous refusals to issue connection conditions that investors receive, long-term grid planning and investments in line with the planned expansion of renewable energy production capacity are necessary, taking into account future demand and the goal of climate neutrality.

3.1.2 ENVIRONMENTAL PROTECTION

In turn, referring to environmental regulations, it should be noted that from the very beginning the requirements in this aspect were very strict. In this case, the administrative procedure for issuing a decision on environmental conditions is important. In light of Article 71 (2) of the Law on Providing Information on the Environment and its Protection, Public Participation in Environmental Protection and Environmental Impact Assessments (Act – Information on Environmental Protection 2008) the implementation of projects that may always have a significant impact on the environment and projects that may potentially have a significant impact on the environment must be preceded by obtaining a decision on environmental conditions. Pursuant to Article 75 (1r) of the EIA Act, the authority competent to issue a decision on environmental conditions for wind farms is the locally competent regional director of environmental protection.

In accordance with Paragraph 2(1)(5) of the Regulation of Council of Ministers on projects likely to have a significant impact on the environment (2019) installations using wind power to generate electricity with a total rated power of not less than 100 MW and located in the maritime areas of the Republic of Poland are among the projects likely to always have a significant impact on the environment (Group I projects), for which an environmental impact assessment is mandatory. In turn, § 3 (1) (6) of the EIA Regulation specifies installations using wind power for electricity generation, other than those listed in § 2 (1) (5), with a total height of not less than 30 m, and located in the areas covered by forms of nature protection referred to in Article 6(1) (1)-(5), (8) and (9) of the Act of April 16, 2004, on nature protection (Law of Nature Protection 2004), i.e., in areas of national parks, nature reserves, landscape parks, protected landscape areas, Natura 2000 areas, ecological land and areas of natural and landscape complexes, included in the projects likely to have a potentially significant impact on the environment (Group II projects), for which the obligation to carry out an environmental impact assessment may be established.

The environmental impact assessment of wind farms is carried out as part of the procedure for issuing a decision on environmental conditions for a planned project that may always have a significant impact on the environment (Group I projects), and may be carried out for a project that may potentially have a significant impact on the environment (Group II projects), if the obligation to carry out the assessment was determined by the competent authority. In addition, the environmental impact assessment procedure may also be carried out as part of the procedure for a decision on a permit for the construction of a wind farm, the so-called re-evaluation of environmental impact, if the authority competent to issue a decision on environmental conditions determines such a need, at the request of the entity planning to undertake the project, or if the authority competent to issue a decision on a construction permit determines that changes have been made in the application for this decision in relation to the requirements specified in the decision on environmental conditions.

As part of the proceedings aimed at issuing a decision on environmental conditions, the direct and indirect impact of a given project on the environment and population, including health and living conditions of people, material goods, monuments, landscape, including cultural landscape, is determined, in accordance with Article 62(1) of the EIA Act, the direct and indirect impact of a given project on the environment and the population, including human health and living conditions, material assets, monuments, landscape, including cultural landscape, the interrelationship between the above elements, the possibilities and ways to prevent and mitigate the negative impact on the environment, as well as the required scope of monitoring, which, in the case of onerous investments, will make it possible to determine the actual impact on the natural and social environment during operation and possibly to apply additional minimizing measures.

The basis for assessing the environmental impact of a project and determining the environmental conditions for the implementation of the project is the report on the environmental impact of the project (the EIA report), the scope of which is directly specified in Article 66 of the EIA Act. It should be noted that for projects in Group

I the applicant shall, as a rule, submit an EIA report immediately. For Group II projects, on the other hand, as a rule, the project information sheet (kip) is submitted first, and after the obligation to conduct an environmental impact assessment is established, the EIA report is submitted.

The purpose of the EIA report is to determine the potential impacts associated with the implementation of the planned project. The purpose of the document in question is to present the impact of a given project on the environment, along with variants and data that would confirm specific impacts. The doctrine emphasizes that the variants of project implementation are one of the most important instruments for the proper assessment of the environmental impact of a project. In turn, the need to present variants of the project, i.e., the variant preferred by the applicant and a rational alternative variant, serves to optimize and rationalize the location, scale and technical and technological solutions of the project. It is worth noting that, according to the EIA Act, it is also necessary to present the rational variant that is most beneficial for the environment, but it follows from court-administrative jurisprudence that the applicant has the right to indicate which variant it considers most beneficial for the environment (judgment of the Provincial Administrative Court in Bydgoszcz of October 5, 2016, ref: II SA/Bd 425/16). In other words, the variant proposed by the investor may coincide with the variant that is most beneficial to the environment (judgment of the Provincial Administrative Court in Olsztyn of July 9, 2020, ref. no.: II SA/Ol 997/19). Thus, there is no need to create a third variant for the implementation of the investment, which would be the most favorable for the environment.

The EIA report should identify the nuisances created during all phases of the investment, i.e., implementation, operation and decommissioning. Then, the authority conducting proceedings to issue a decision on environmental conditions shall thoroughly verify the data presented in the environmental documentation. The authority conducting administrative proceedings, on the basis of the evidence, should include in the decision the conditions the fulfillment of which will guarantee that the implementation and operation of the project will not significantly affect the environment, including human health. It is worth emphasizing that *the authority issuing this decision* [decision on environmental conditions] *should carry out the proceedings provided for by the provisions of the aforementioned law* [the EIA Act] *and is obliged to issue this decision if the investor meets the requirements set forth in the provisions of the Act* (judgment of the Provincial Administrative Court in Rzeszów of October 18, 2017, ref: II SA/Rz 861/17). Thus, in principle, the refusal to establish environmental conditions for wind farms for which an environmental impact assessment has been carried out, in the form specified in the application, may take place in statutorily defined cases, i.e.,

- Non-compliance of the location of the project with the provisions of the local spatial development plan (Article 80(2) of the EIA Act);
- Lack of possibility to implement the project in variants referred to in Art. 66 (1) (5) of the Environmental Protection Act and in case of lack of consent of the applicant to indicate in the decision on environmental conditions the variant allowed for implementation (Art. 81 (1) of the Environmental Protection Act);

- demonstration of a significant negative impact on a Natura 2000 area, unless the prerequisites referred to in Art. 34 of the Law on Nature Conservation occur (Art. 81(2) of the Law on Environmental Protection);
- demonstration that the project may adversely affect the possibility of achieving the environmental objectives referred to in Article 56, Article 57, Article 59 and Article 61 of the Law of July 20, 2017 – Water Law (Article 81(3) of the EIA Law);
- failure to maintain the minimum distance of a wind farm from a residential building or a building with a mixed function that includes a residential function, or selected forms of nature conservation (Article 6(7) of the Law of May 20, 2016, on wind power investments (Distance Act 2016)).

An important element of the procedure for issuing a decision on environmental conditions is to obtain the opinions and agreements required by the law. Thus, if an assessment of the environmental impact of a project is carried out before issuing a decision on environmental conditions, the authority issuing this decision agrees with the conditions for implementation of the project with the authority competent for water assessments, as referred to in the provisions of the Act of July 20, 2017 – Water Law – and consults the authority of the State Sanitary Inspectorate. At this stage, moreover, a substantive inspection of the content of the environmental documentation is carried out, the purpose of which is to determine the conditions for implementation and operation of the project. The authority conducting the proceedings for the issuance of a decision on environmental conditions assesses whether the proposed location in a specific area complies with the provisions of the local spatial development plan, as well as whether it does not threaten areas of natural value, including Natura 2000 areas. It is worth pointing out at this point that according to Article 4(2) of the Distance Law, when locating and constructing wind facilities, it is necessary to maintain a distance equal to or greater than ten times the height of the wind farm from the forms of nature protection referred to in Article 6(1) points 1–3 and 5 of the Law of April 16, 2004, on nature protection, and from the forest promotion complexes referred to in Article 13b(1) of the Law of September 28, 1991, on forests (Act on Forests 1991). One should also bear in mind here Article 33 (1) of the Law on Nature Protection, according to which it is forbidden to undertake activities that, separately or in combination with other activities, may have a significant negative impact on the objectives of protection of Natura 2000 area, including, in particular, worsening the condition of natural habitats, habitats of plant and animal species for the protection of which Natura 2000 area was designated or adversely affect the integrity of Natura 2000 area or its connections with other areas. If it is determined that Article 33(1) of the Law on Nature Protection applies and that the grounds for an exception to this prohibition, as specified in Article 34(1) or (2) of the Law on Nature Protection, do not exist, the locally competent regional director of environmental protection shall deny permission for the implementation of the wind farm.

In the issued decision, the concurring authority determines, among other things, ways to reduce impacts at the stage of implementation and operation (e.g., indicates the period of construction works, solutions in water and sewage management, technical and technological solutions aimed at minimizing or eliminating the impact of the project on the environment). If it is not possible to reduce or exclude the above-normative impact on the environment, the authority competent for water law assessments refuses to agree on the conditions for implementation of the project. The opinion of the State Sanitary Inspection authority, on the other hand, refers to the hygiene and health requirements that should be taken into account in the implementation of the project.

It should be emphasized that the content of the decision on environmental conditions, issued after the environmental impact assessment of a project, follows directly from Article 82 of the EIA Act. In Article 82 (1) (2), the legislator regulated the situations whose occurrence requires making certain determinations in the decision on environmental conditions. In the case of a project for which an environmental impact assessment has been carried out, the decision may specify the need to perform environmental compensation, as well as the obligation to prevent, reduce and monitor the impact of the project on the environment. Thus, if the environmental impact assessment shows the necessity to impose the aforementioned elements, then the locally competent regional director of environmental protection is obliged to determine them in the decision concluding the administrative proceedings. At the same time, the legislator leaves it up to the public administration body to freely impose the obligation to present a post-implementation analysis, specifying its scope and the deadline for its presentation. The legislator does not refer to any prerequisites on which it makes the imposition of such an obligation (Rakoczy 2010).

This analysis is intended to show the actual nuisance to neighboring properties. It is appropriately prepared in the interest of their owners (judgment of the Supreme Administrative Court in Warsaw of December 8, 2011, ref: II OSK 2169/11). It is necessary to indicate in detail to what extent and by what date it should be prepared. The decision on environmental conditions should specify the authorities to which the prepared information should be presented. The post-implementation analysis compares the findings contained in the EIA report and the decision on environmental conditions with the actual environmental impact of the project and the actions taken to mitigate it. This is primarily about actions regarding the anticipated nature and extent of environmental impact, as well as planned preventive measures. It should be noted, however, that the key difference between an EIA report and a de facto post-implementation analysis is the deadline by which the aforementioned documents should be prepared.

The legislator stipulates that when the post-implementation analysis shows that it is necessary for the project to establish a restricted use area (causing restrictions on the use of real estate), the analysis should be accompanied by a copy of the cadastral map certified by the competent authority with the course of the boundaries of the area marked.

It should be borne in mind that the investor, obliged in the decision on environmental conditions to submit a post-implementation analysis by a given date, should comply with this obligation. Otherwise, it is possible to restrict the use of the property that is the subject of their investment. An example of this would be if the provincial environmental inspector were to suspend the commissioning of a constructed/ rebuilt facility or set of facilities related to the project.

Simultaneously, it is worth noting that regardless of the provisions of the decision on environmental conditions, a wind farm, according to Article 76 of the Environmental Protection Law, cannot be put into operation if it does not meet the environmental protection requirements specified in this provision.

An important factor to be analyzed that constitutes the nuisance during the operation of wind farms corresponds to acoustic impacts. With regard to the areas designated for residential buildings, hospitals and social care homes, buildings associated with permanent or temporary residence of children and young people, as well as for spa, recreational and leisure, residential and service purposes, the minister responsible for the environment, in consultation with the minister responsible for health, established, by way of a decree of the Minister of Environment of June 14, 2007, on permissible levels of noise in the environment (Regulation – noise 2007), expressed by $L_{Aeq\,D}$ and $L_{Aeq\,N}$ indicators. If the environmental impact assessment of the project reveals the possibility of exceeding acoustic standards, it will be necessary to take measures to minimize the impact to a level that will not adversely affect the functioning of people living in the vicinity of the proposed wind turbine.

It is also important to note the situation when the authority conducting the proceedings for the issuance of a decision on environmental conditions determines that there is no need to assess the environmental impact of a wind turbine (possible only for a project of the second group – Paragraph 3(1)(6) of the EIA Regulation). In such a situation, the applicant is not required to submit an EIA report, while the project information sheet constitutes the basis for determining the conditions for the implementation and operation of the project. The scope of this document is strictly defined in Article 62a of the EIA Act. Simultaneously, it should be emphasized that, as a rule, refusal to determine the environmental conditions for a wind turbine for which no environmental impact assessment has been carried out, in the form specified in the application, may take place in statutorily indicated cases, i.e.,

- Non-compliance of the location of the project with the provisions of the local spatial development plan (Article 80 paragraph 2 of the EIA Act);
  failure to maintain the minimum distance of a wind turbine from a residential building or a building with a mixed function that includes a residential function or selected forms of nature conservation (Article 6(7) of the Distance Act (2016)).

In the decision on environmental conditions for a wind farm for which an environmental impact assessment has not been carried out, the locally competent regional director of environmental protection may specify:

- important conditions for the use of the environment in the implementation and operation or use phase of the project, with particular regard to the need to protect valuable natural values, natural resources and monuments, as well as to limit the nuisance to neighboring areas;
- environmental protection requirements necessary to take into account in the documentation required to obtain decisions, issued under the Act of July 7 (Construction Law 1994);

- as well as impose an obligation to avoid, prevent, reduce the impact of the project on the environment or monitor the environmental impact of the wind farm.

According to Section III of the EIA Act, "Public Participation in Environmental Protection," the competent regional director of environmental protection shall, prior to issuing a decision on environmental conditions, ensure that the public has the opportunity to participate in the proceedings in which the environmental impact assessment of the project is carried out. It is worth mentioning at this point the rights of an environmental organization (Article 44(2) of the EIA Act), including the right to appeal against a decision issued in proceedings requiring public participation, even if it did not participate in those proceedings before the first instance authority, provided that the organization has been carrying out statutory activities in the field of environmental protection or nature protection for a minimum of 12 months prior to the date of initiation of those proceedings. The accession of an environmental organization to the proceedings aimed at issuing a decision on environmental conditions as a party then occurs by virtue of a declaration of intent made by the organization. It is not necessary to issue a relevant decision in this regard. It is also important to point out that it follows from Article 44(1) of the EIA Act that an environmental organization may declare its willingness to participate in a particular proceeding requiring public participation at any stage of the proceeding. The situation is different for associations that do not meet the abovementioned statutory prerequisites under the EIA Act. Then, the association may join the administrative proceedings in accordance with Article 31(1) of the Act of June 14, 1960 (Code of Administrative Procedure 2021), as a social organization. The basis for issuing a decision on admitting this association to the proceedings as a party is the fulfillment of two prerequisites, namely, it is justified by the statutory objectives of this organization and the public interest behind its participation. At the same time, it should be emphasized that only joining the proceedings as a party in the first instance entitles a social organization to file an appeal against the decision on environmental conditions, and then to file a complaint to the administrative court against the decision of the General Director of Environmental Protection, i.e., the decision issued in the appeal proceedings.

　　Here, it is necessary to emphasize that, regardless of the provisions of the EIA Act, a party to the proceedings referred to in Article 74(3a) of the EIA Act, in the proceedings aimed at issuing a decision on environmental conditions, has the right to active participation at every stage of the proceedings. This right follows directly from Article 10(1) of the Code of Administrative Procedure. Moreover, the locally competent regional director of environmental protection is obliged to allow the parties to the proceedings to comment on the evidence and materials collected before issuing a decision on environmental conditions.

3.1.3　Planning

Already in 2018, Sudra and Bida-Wawryniuk (2018) distinguished numerous principles characterizing Polish legislation on environmental impact assessment and requirements for the location of wind turbines, which, in addition to the infamous rule under Article 4(1) of the Distance Law (Rule 10H), influenced the formation of distances from built-up areas and areas planned for development by residential construction or with a residential function (and an analogous requirement for development in relation to existing or planned power plants in the local plan). Statutory restrictions can now include:

- the need to maintain a minimum distance from selected forms of nature conservation, in accordance with the provisions of the Distance Law;
- the possibility of establishing a wind farm in Natura 2000 areas, only if the environmental impact assessment does not indicate a significantly negative impact of the investment on these areas, or obtaining derogations from Article 34 (1) of the Law on Nature Protection – ensuring the performance of natural compensation necessary to ensure the coherence and proper functioning of the Natura 2000 network;
- allowing the establishing of a wind farm in a landscape park only if the environmental impact assessment procedure carried out has shown that there is no adverse impact on the nature and landscape of the landscape park (Article 17 (3) of the Law on Nature Protection), while national parks and nature reserves are excluded from the location (Article 15 of the Law on Nature Protection);
- expanding the scope of landscape impact assessment as part of the prepared EIA report, which was introduced with the Act of April 24, 2015, amending certain acts in connection with the strengthening of landscape protection tools (Landscape Law 2015), which amended, among other things, the EIA Act introducing the obligation to perform an analysis and assessment of the impact of the planned investment on the landscape, including the cultural landscape (Article 62(1)(1)(ca) of the EIA Act). The abovementioned provisions led to the extension of the necessary scope of the EIA report (Article 66(1)) to include a description of the landscape in which the investment is to be located and a justification of the selected variant of the project in terms of its landscape impact;
- protection of agricultural land of high valuation classes I–III, because according to Article 7(2) of the Act of February 3, 1995, the protection of agricultural and forest land (Act – Protection of Agricultural Land 1995) limits the use of land of classes I–III for non-agricultural and non-forest purposes. Whenever the construction of wind turbines is to take place on land that is class I–III agricultural land, the investment must be implemented on the basis of a local land use plan and it is mandatory to obtain the consent of the minister responsible for rural development;
- the need to maintain a minimum distance equal to ten times the total height of the wind farm when locating these facilities in relation to forest promotion complexes, as referred to in Article 13b of the Act of September 28, 1991, on forests (Act on Forests 1991);
- the need to obtain a water law permit for the location of a wind farm in areas at special risk of flooding, in accordance with Article 390(1)(1)(a) of the Act of July 20, 2017 (Water Law 2017);

- prohibition of the construction of wind farms in zone "A" of spa protection, in accordance with Article 38a (1j) of the Act of July 28, 2005, on spa treatment, spas and areas of spa protection and spa municipalities (SPA protection Act 2005);
- maintaining the distance from the outer edge of the roadway of construction objects, including wind facilities, in accordance with Article 43 (1) of the Act of March 21, 1985, on public roads (Act of Public Road 1985). The required distance depends on the type of road (highway, expressway, national road, provincial road, county road, municipal road) and the category of land (development area, outside development area);
- legal and planning protection of historical monuments, in accordance with the Law on the Protection and Care of Historical Monuments of July 23, 2003 (Protection Monuments Act 2003);
- agreeing the location with the president of the Civil Aviation Authority by issuing an opinion pursuant to Article 877(2) of the Act of July 3, 2002 (Aviation Law 2002) – on the possibility of locating the investment due to the height of the development in the area where the areas limiting development are in force.

Therefore, it should be pointed out that when selecting the location of wind farms, the applicant must take into account numerous legal objections. Land use planning issues should also be closely connected with investment location planning. On the basis of the generally applicable regulations on aspects of planning, construction, impact assessment, technical requirements, grid operation, support system, etc., wind farms have been among the most thoroughly studied projects, even before obtaining construction and operating permits. The industry's rapid growth in 2010–2011 was further complicated by additional environmental requirements stemming from specialized guidelines (discussed in detail later in this publication), most of which were in informal use, although not all were accepted and officially adopted. Between 2010 and 2016, certain rules regarding the implementation of wind investments changed and new regulations emerged, placing additional restrictions on the industry.

The Law on Spatial Planning and Development of March 27, 2003 (Spatial Planning Law 2003), defines the basic principles for the formation of spatial policy by local government units and government administration bodies. The normative act in question defines the scope and methods of proceeding in matters of allocating land for specific purposes and determining the principles of its development and construction, taking spatial order and sustainable development as the basis for these activities.

The structure of the system is systematized and divided into three levels, i.e., national, regional and local. At the national level, the minister responsible for construction, planning and spatial development and housing, taking into account the principles of sustainable development as well as the goals and directions of the country's long-term development strategy, prepares the Concept for the Spatial Management of the Country. The last such document (Concept for Spatial Development of the Country 2030) was adopted by Resolution No. 239 of the Council of Ministers on December 13, 2011, which came into force on April 27, 2012. The act repealing the aforementioned resolution was the Act of July 15, 2020, amending the Act on the principles of development policy and certain other acts (Development Policy Act 2020). Nevertheless, the provisions of the aforementioned concept should be introduced into provincial spatial development plans, which is also a document at a regional level. The provincial spatial development plan combines national and local planning. It shapes the spatial policy of the province, in accordance with the Concept of National Spatial Planning, taking into account the territorialization of development policy. The plan formulates the objectives of management of the provincial space and the principles of its formation, and determines the directions of spatial policy in the long term. It provides the basis for the construction of operational programs for the development of the province. It should be noted that the plan does not constitute a local law, but is an act of internal management and binds the provincial authorities, as well as other public administration entities, to respect development priorities and conduct spatial policy in accordance with the established directions. With regard to the location of wind farms, the provincial spatial development plan can only roughly indicate the location of investments, but the planning authority in this regard is the municipality.

Spatial planning at the local level is based on:

- a study of the conditions and directions of spatial development of the municipality;
- local spatial development plan;
- decisions on the conditions of development and land use;
- decisions on the location of public purpose investments.

In the context of the implementation of wind power investments, only the first two planning forms are relevant. This is due to the provisions of Article 3 of the Distance Law, where it is explicitly indicated that *a wind farm is established only on the basis of a local zoning plan*. Thus, wind turbines cannot be located on the basis of a decision on land development conditions. Furthermore, in accordance with the doctrine (Zajdler 2012) wind facilities do not constitute a public purpose investment; therefore, a decision on determining the location of a public purpose investment cannot be the basis for their implementation.

The study determines the spatial policy of the municipality and the principles of spatial development, taking into account the principles set forth in the Concept of National Spatial Development, the findings of the development strategy and the spatial development plan of the province, the framework study of the conditions and directions of spatial development of the metropolitan association as well as the development strategy of the municipality (in case the municipality has such a study). It forms the basis for all activities undertaken in the municipality in the field of planning and spatial development. Its findings are binding for municipal authorities when drawing up local spatial development plans. The study is a document on the basis of which the municipality can formulate proposals for the spatial development plan of a municipality and verify the arrangements adopted in it for its area. The study is adopted by the municipal council in order to determine the spatial policy of a municipality, including local zoning rules (Article 9 (1) of the Local Spatial Development Law). The thematic scope of the study includes, but is not limited to (Art. 10 (1) and (2) of the Local Spatial Development Law):

- issues related to the development of the municipality, including those arising from the identified existing development, land ownership, quality of life
  of residents, tasks for the implementation of supra-local public objectives,
- protection areas of natural and cultural environment and the principles of
  their protection and use,
- directions of changes in spatial structure and land use, principles of development of communication and technical infrastructure systems,
- principles of municipal policy in the preparation of local plans.

It is worth noting that the provision of Article 10 (2a) of the Local Spatial Development Law stipulates that if in the municipal area it is envisaged to designate areas where devices generating energy from renewable energy sources with a capacity exceeding 100 kW will be deployed, as well as their protection zones related to restrictions on development and land use and development, then the study shall determine their approximate location.

A local plan is drawn up to determine the use of land, including for public purpose investments, and to specify the ways in which it may be used and developed (Article 15 (1) of the Local Spatial Development Law). The local plan specifies, i.e.:

- principles of development and the ratio and intensities of integration and
  division of real estate, modernization, development and construction of communication systems and technical infrastructure, as well as rules to protect
  the environment, nature, landscape and cultural heritage and monuments,
- principles of protection and shaping of spatial order, principles of protection of the environment, nature and landscape with guidelines for landscape
  shaping, or principles of protection of cultural heritage and monuments,
  including cultural landscapes, and modern cultural assets,
- boundaries of areas of high flood risk, priority landscapes defined in the
  landscape audit and in the province's spatial development plans.

Article 14(8) of the Local Spatial Development Law stipulates that a local zoning plan is not only a planning act but also an act of local law. Local law acts, in accordance with Article 87(2) of the Polish Constitution, are sources of universally applicable law in the area of activities of the authorities that established them. Therefore, the provisions in the area covered by a particular local zoning plan have universally binding force. The local plan is a concretization of the concepts contained in the study of conditions and directions for spatial development, but only it can be the basis for issuing decisions related to planning and spatial development. It is important that the content of the local plan must not contradict the content of the study of the conditions and directions for spatial development of the municipality.

In accordance with the procedure for adopting a study of municipal conditions and directions for spatial development and local development plans, it is required to ensure public participation in the process. Pursuant to Article 17 of the Local Spatial Development Law, after the municipal council adopts a resolution to proceed with the preparation of a local plan, the head of the municipality (mayor or president of the city) publicly announces this fact along with information about the possibility of submitting applications to the plan. It is necessary to specify the form, place and deadline for submitting applications. The comments submitted during consultations are subject to consideration in the relevant procedure. A public discussion is also held on the solutions adopted in the draft local plan and the possibility of submitting comments on the draft plan. The local plan as an act of local law is also subject to judicial and administrative control. Finally, the head of the municipality (mayor, president of the city) presents the draft local plan to the municipal council along with a list of the comments that have not been taken into account.

In terms of wind energy, highly important is Article 15(3)(3a) of the Local Spatial Development Law, which indicates that a local spatial development plan shall specify the boundaries of areas for the construction of the devices referred to in Article 10(2a) of the Local Spatial Development Law discussed when analyzing the legal analysis of the municipality's spatial development study (i.e., devices generating energy from renewable energy sources with a capacity exceeding 100 kW) and the boundaries of their protection zones related to restrictions on building, development and use of land and the occurrence of significant environmental impacts of these devices.

A wind farm is a construction object within the meaning of the Construction Law, which may be built only after obtaining a construction permit. A construction permit is understood as an administrative decision authorizing the commencement and conduct of construction or the performance of construction works other than the construction of a building object (Article 3(12) of the Construction Law). The construction permit decision is granted at the request of the investor. According to Article 33(2) of the Construction Law, an application for a construction permit must be accompanied by, among other things: a construction project, a statement of the right to dispose of the real estate for construction purposes, or the previously discussed decision on environmental conditions. Pursuant to Article 35 of the Construction Law, before issuing a construction permit or a separate decision to approve a construction project, the competent authority shall verify:

- compliance of the construction project with the provisions of the local spatial development plan or the decision on the conditions of development and land use in the absence of a local plan, as well as the requirements of environmental protection, in particular, those set forth in the decision on environmental conditions;
- compliance of the plot or land development project with regulations, including technical and construction regulations;
- completeness of the construction project and possession of the required opinions, agreements, permits and verifications, as well as the information on safety and health protection referred to in Article 20(1)(1b) of the Construction Law, the certificate referred to in Article 12(7) of the Construction Law and the documents referred to in Article 33(2)(6) of the Construction Law;
- verification of the project referred to in Article 20(2) of the Construction Law by a person having the required construction authorization and holding a certificate referred to in Article 12(7) of the Construction Law, valid as of the date of its verification;
- the conditions for connection to the grid in view of the contents of Article 34(3a) of the Construction Law.

10H
Excluded: 99.72%
Available: 0.28%

Nature Protection
Excluded: 92.92%
Available: 7.08%

FIGURE 3.1   Availability of land for wind investment before and after liberalization of the 10H Rule (Hajto et al. 2017).

In the context of planning, the most relevant, in the opinion of the authors of this publication, is the restriction resulting from the rule set forth in Article 4(1) of the Distance Law (Rule 10H), which excludes 99.7% of the area of Poland from wind investments (Czyżak et al. 2021). Figure 3.1 shows that the liberalization of the 10H rule as resulting from the "Nature Protection" scenario allows to increase the availability of land to 7.08% – more than 25 times. At the same time, it is worth noting that attractive areas in the north of the country, particularly in the Pomeranian and West Pomeranian provinces, can be unlocked.

This is also confirmed by a comprehensive study of the location considerations for wind farms conducted by the IOŚ-PIB Team, including modeling of exclusions resulting from maintaining a buffer of 1 and 2 km from development. The adopted buffers exclude from the possibility of investing in wind farms 93.9% (1 km buffer) and 99.1% (2 km buffer) of Poland's area, respectively, so they are basically equivalent to a ban on localization (Hajto et al. 2017).

3.1.4   TAX ISSUES

The main tax burden on a producer of energy from a wind farm located onshore, besides income tax, is property tax. *Property tax is levied on land, buildings or parts thereof and structures or parts thereof related to the conduct of business activities. The amount of liability depends on the subject of taxation* (Polish Wind Energy Association 2022). The Law of January 12, 1991, on Local Taxes and Fees (Tax Act 2019) sets maximum property tax rates in all categories, but local government units have the right to set an individual rate in their area, but they cannot be higher than the statutory rate. In the case of onshore wind turbines, only the structural parts of the wind turbines, namely the foundation with the ring and the tower, are subject to property tax. The other components of the turbine unit are generally subject to exclusion from this tax. It should be noted that in 2017 there was a temporary change in the rules for accounting for property tax and subjecting the entire value of a wind turbine to taxation. However, at the beginning of 2018, the original wording of the rules was restored and remains in effect today. In practice, there are noticeable differences in the effective property tax burden on farms. They are mainly due to technological differences in tower construction and installation, and sometimes to different methods of segregating investment costs at the stage of commissioning the project. The legal definition of a wind farm was introduced by the Distance Law. According to Article 2(1) of the indicated normative act, a wind farm consists of a building part, which is a structure within the meaning of the Construction Law, and technical equipment, including technical elements, in which electricity is generated from wind energy As of January 1, 2018, the renewable energy industry has recognized that a wind farm is a structure. That is why the tax on wind turbines is levied only on the construction parts, which account for about 30% of the value of the entire farm. The tax on structures in Poland amounts to 2% of the value of the structure, determined in accordance with Article 4(1)(3) and (3–7) of the Tax Law.

Concerns were raised by Article 17(2) of the Law of June 7, 2018, amending the Law on Renewable Energy Sources and certain other laws (Renewable Energy Sources Act 2021), which provided for the entry into force, with retroactive effect from January 1, 2018, of amendments to the provisions of the Construction Law, the Annex to this Law and the Law on Wind Power Plant Investments, affecting the scope of property taxation. These changes have led to a reduction in property tax revenues for local governments, particularly for small municipalities where wind farms are located.

This is because before the amendment, starting in 2017, property tax was charged on the entirety of the equipment, while after the 2018 amendments, the tax was to be paid only on the construction part of the wind turbines. The disputed legislation was enacted in mid-2018, but the regulations were retroactive to January 1, 2018. Determining the retroactive entry into force of the regulations resulted in municipalities being forced to refund taxpayers for the tax they had previously paid.

The changes met with rather fierce opposition from the municipalities, which petitioned the Constitutional Court to examine the compatibility of the amendments with the Polish Constitution. This is because the amendment violated the principle of non-retroactivity of laws (lex retro non agit). According to the applicants, while exceptions to this principle are possible, it is absolutely impermissible to legislate retroactively to make the law more severe, which was the case here. In addition, the municipalities argued that the regulations introduced are incompatible with the constitutional principle of self-government independence, which includes the principle of financial independence (Jagoda 2011). *The reduction of the municipalities' income from the real estate tax has caused problems with covering the anticipated expenses, primarily related to the implementation of own tasks, and the applicants have shown that this leads to a deterioration in the quality of life of their residents, especially in small municipalities where wind farms are located* (Chojnacka & Nowicki 2020). The Constitutional Court found that the legislator retroactively introduced provisions strictly related to tax law in a situation where this was not necessary, without observing the principle of proportionality and worsening the financial situation of municipalities where wind farms are located.

The Constitutional Court, in its judgment of July 22, 2020, ref. K 4/19 (Journal of Laws 2020.1336), found Article 17(2), to the extent that it retroactively introduced Article 2(1) and (6) and Article 3(1) of this amending law, to be inconsistent with the principle of non-retroactivity of law derived from Article 2 of the Polish Constitution. The Constitutional Court noted that due to the retroactive amendment of the definitions of "structure" and "wind farm," municipalities were deprived of part of their revenue and, in addition, obliged to refund overpaid property tax collected from wind farms located on their territory for the period from January to June 2018.

In accordance with the judgment of the Constitutional Court in question, the disputed regulation ceased to be effective 18 months after the judgment was published in the Journal of Laws. Until that time, the challenged regulation remains in force and will be applied. This is because the Constitutional Tribunal considered it justified to minimize the impact of the issued judgment on already formed legal relations and decided that it was necessary to postpone the date on which the regulation inconsistent with the Constitution ceased to be in force, indicating at the same time that it is aware that this is not a sufficient instrument to ensure the protection of property rights of municipalities. During the period of deferral of the expiration of the validity of the regulation, the legislator was obliged to bring the regulations in the disputed area into conformity with the Constitution of the Republic of Poland. The realization of the above was ensured by the Act of November 17, 2021, on compensation of income lost by municipalities in 2018 in connection with the change in the scope of taxation of wind farms (Act of taxation of wind power plants 2021), which enabled compensation for losses arising in local government budgets.

According to the law, reimbursement is made upon application submitted by the municipality to the governor, in which the local government should indicate, among other things, the amount of revenue lost in 2018 due to the change in the law, but also the list of wind farms located in the municipality and the tax base of the structures included in these plants.

Another important issue in the context of the issue at hand is the depreciation of wind farms. The expenses incurred in the construction of a wind farm are tax deductible through depreciation allowances. In Poland, the area of tax depreciation in Poland forces companies, and regardless of the economic sector, to keep separate depreciation tables for tax and balance sheet purposes. Tax depreciation rates are limited by law (Polish Wind Power Industry 4.0 2022). From the plant's capital expenditures, individual components are separated and depreciated at the appropriate rates. Towers, platforms and foundations of wind farms are classified in group 201 of the Classification of Fixed Assets (CFA) as "structures on wind farm sites" and depreciated at the rate of 4.5%. Technical parts are classified in group 346 of the CFA "wind power generating sets" and should be depreciated according to the straight-line method at a rate of 7% or degressive with the application of a factor of 2.0 (a rate of 14%). It should also not be overlooked that fixed assets of wind farms comprise medium- and high-voltage power cables, as well as the connection to the grid. These fixed assets are included in Group 2 of the CFA (CFA 211), the linear rate is 10%. *However, it is worth pointing out that, according to the appendix to the Corporate Income Tax Law, the 10% rate is intended for assets in group CFA 211, but only those under the name "Wires of in-house technological networks." In other cases, the rate of 4.5% applies. The final qualification of an investment in a connection to the distribution or transmission network of the DSO/TSO depends on a number of variables, including detailed design arrangements, the point of connection, i.e., the property boundary between the investor's and the operator's infrastructure, the method of accounting for the investment agreed with the operator, etc. In terms of depreciation of access and technical roads (CFA 220), a linear depreciation rate of 4.5% is assumed* (Polish Wind Power Industry 4.0 2022). In addition to the above-mentioned elements of the investment, there are numerous other elements, the qualification of which is already made by the taxpayer himself, who, in case of doubt, may ask a competent statistical authority for assistance (Central Statistical Office – CSO). Confirmations of the grouping of CFA from the CSO are considered binding on the subject of determining the appropriate depreciation rate.

An entrepreneur who purchases wind turbines together with their installation from a foreign entity, not registered in Poland for VAT purposes, is obliged to tax such transaction. The place of taxation of the goods corresponds to the place of their installation. Thus, it is always necessary to verify whether the foreign entity has registered as a VAT taxpayer or not, although it was obliged to do so, in connection with, for example, a permanent place of business in Poland. Such verification will allow avoiding the situation of the so-called reverse charge, when a foreign unregistered entity supplies goods. In such a case, the obligation to account for VAT rests with the buyer (investor).

The final aspect relating to the issues at hand is the taxation of long-term corporate power purchase agreements (cPPAs) between a power generator and an end user. Despite the numerous benefits of cPPAs, there are also some doubts in this matter, including, above all, the tax aspect. This issue may concern, i.e., the qualification of virtual power purchase agreements (vPPAs) based on the structure of a differential contract (settlement is made by settling the differences between the contracted price and the current market price) against the provisions of the Law of March 11, 2004, on tax on goods and services (VAT Act 2004). Relative to the current practice of applying the VAT law, the vPPA constitutes the provision of services related to financial instruments, and not the supply of energy. Thus, there is no basis here for using the subject exemption from VAT. However, it should be pointed out that the VAT law does not regulate the cases when *an energy generator is obliged to pay a monthly settlement amount to the purchaser when, with respect to the contracted energy, the market price turns out to be lower, and whether, as a result, the purchaser provides the generator with a service that is covered by VAT regulations or outside their scope* (Polish Wind Power Industry 4.0 2022).

The second major limitation arising from cPPAs is excise tax obligations under Article 9 of the Excise Tax Law of December 6, 2008 (Excise Act 2008). The law provides for an exemption from excise duties for energy produced from a renewable energy source, but this involves the acquisition of green certificates (as discussed earlier in this publication). Therefore, in the case of an exemption from the obligation to redeem certificates of origin for electricity sold under the cPPA formula, the exemption referred to in Article 30(1) of the Excise Law does not apply. In the case of cPPAs, the power generator is required to pay excise taxes by the 25th day of the month following the month in which the differential settlement for the period in question was made. According to Article 89(3) of the Excise Law, the excise tax rate for electricity is PLN 5/MWh.

3.2   STRATEGIC DOCUMENTS

Wind farms, like all other types of projects, are implemented in a certain legal environment, understood at the most general level as the implementation of EU and national policies, programs and strategies. At the level of specific regulations – as projects – they are subject to the regulations on the investment process, which is discussed in detail in Section 4.1. In the case of wind energy facilities, they are further implemented based on the content of strategic documents. The most important documents relating to the conditions for wind energy development include those indicated below:

3.2.1 POLAND'S ENERGY POLICY UNTIL 2040 (PEP2040)

The document shows that a zero-carbon system is a long-term direction, and assumes the transitional use of gas in industry, greater use of offshore wind turbines and increasing the role of so-called distributed generation, consisting of the production of electricity by small, domestic generating units. However, it should not be overlooked that while the document in question includes issues related to the development of offshore wind farms, the document ignores the development of onshore wind farms. PEP2040 implies that the share of energy generated from renewable energy sources in the electric power industry will be at least 32% net in 2030 and 40% net in 2040. However, it cannot be overlooked that the assumptions made in the document, the boundary costs, including CO_2 emission rates, have changed significantly since the drafting of PEP2040. It is therefore necessary to revise the strategic document in question. Without revising PEP2040, it will be impossible to further develop renewable energy sources in Poland. The above is due, for example, to the fact that the adopted level for the share of renewable energy sources in the country's energy mix at 32% net in 2030 had already been almost reached in 2021 (30.3% of installed capacity).

3.2.2 PEP2040 UPDATE

In early 2022 the Council of Ministers adopted the assumptions for the PEP2040 Update – Strengthening Energy Security and Independence, submitted by the Minister of Climate and Environment. The purpose of this update is to neutralize or reduce the risks associated with potential national and international crises. The current international situation, which has drastically changed as a result of Russia's invasion of Ukraine, affects many aspects related to energy policy. The updated PEP2040 must take into account Poland's energy sovereignty first and foremost. It is about diversification of supply, investments in production capacity, linear infrastructure and storage, and alternative fuels. From the point of view of the wind industry, the assumptions of the PEP2040 Update adopted by the Council of Ministers assume further development of renewable energy sources, with an emphasis on the fact that by 2040, half of electricity generation is to come from renewable sources.

3.2.3 NATIONAL ENERGY AND CLIMATE PLAN 2021–2030 (NECP)

NECPs are strategic documents that all European Union member states have been required to develop. They present a vision for the development of the entire fuel and energy sector, together with an assessment of the impact on the economy, environment and society in the perspective of year 2030 (which may be extended to 2040).

According to the assumptions, Poland will focus on diversification of energy carriers, gradually increasing the share of renewable energy sources (the role of which in the electric power sector will be increased mainly through two technologies, i.e., wind power and photovoltaics), as well as introducing nuclear power into the energy balance starting from 2033. According to the NECP, the share of renewable energy sources in the country's energy mix is expected to gradually increase – from 18% in 2015 to about 40% in 2030 and 50% in 2040 (66% will correspond to wind power).

3.2.4   Poland's Environmental Policy 2030 (PEP2030)

The role of PEP2030 is to ensure Poland's environmental security and a high quality of life for all residents. The policy reinforces the government's efforts to build an innovative economy with sustainable development principles. Its main objective is to develop the potential of the environment for the benefit of citizens and entrepreneurs, which corresponds directly to the objective from the "Environment" area in the Strategy for Responsible Development. PEP2030 presents practical solutions for specific directions of intervention. The specific objectives relate to health, economy and climate. The implementation of environmental goals is to be supported by increasing the effectiveness of environmental instruments. The idea is to develop the public's environmental competence, skills and attitudes, and to improve the management of environmental protection in Poland.

3.2.5   National Reconstruction Plan (NRP)

The strategic goal of NRP is to rebuild the development potential of the economy lost as a result of the COVID-19 pandemic, including building sustainable economic competitiveness. It stems from the European Reconstruction Instrument and includes Energy among the main areas. Poland is to be the beneficiary of a pool of nearly €58 billion, with €23.1 billion in non-refundable grants and €34 billion in low-interest loans. Importantly, the NRP assumes support for the development of distributed energy from renewable energy sources in the form of an amendment to the Distance Law, which will ensure the development of onshore wind farm investments in municipalities that are willing to locate such infrastructure.

3.2.6   National Renewable Energy Action Plan

The National Renewable Energy Action Plan was prepared on the basis of the scheme prepared by the European Commission (Commission Decision 2009/548/EC of June 30, 2009, establishing a scheme for National Renewable Energy Action Plans under Directive 2009/28/EC of the European Parliament and of the Council). However, the document seems to be obsolete, as it assumes a path to a min. 15% share of energy from renewable sources in final energy consumption by three sectors: electricity, heating and cooling, and transport (where the share of energy from renewable sources should be min. 10%).

The site selection procedure, although often very complex due to the need to take into account many factors, is usually not problematic. However, social and political factors (Kolendow 2016), understood as the acceptance and attitude to investment of the local community and its representation in the form of local government units, particularly at the municipal level, remain a challenge. Many regions, recognizing the role and challenges facing the renewable energy industry, have adopted regional documents – wind investment location studies or dedicated to all renewable sources. These documents, on the one hand, provided local authorities with guidance on the preliminary assessment of the suitability of their land for such investments. On the other hand, they fulfilled the regions' obligations to support the design of priorities and disbursement of funds in the Regional Operational Programs (ROPs) with long-term program documents. As part of this work, a number of regional studies were produced, which were expressions of the individual characteristics of the regions and their natural conditions, as well as the energy policies implemented by the regions. Attempts to comprehensively address the criteria for establishing wind farms individually or among other RES in Poland at the regional level were made in the Lower Silesia (Zathey 2010), Lublin (Maleńczuk 2009), Opole (Badora 2010), Pomeranian (Niecikowski & Kistowski 2008), Subcarpathian, or Warmian-Masurian (Instytut OZE 2013) provinces.

3.3   GUIDELINES FOR WIND POWER INVESTMENT PROCESSES

The most restrictive element in the investment process with respect to wind power investments are environmental issues. At the end of 2010, important and first such detailed guidelines for forecasting the environmental impact of wind farms were made public, developed for the European Union countries by the European Commission (2011) authorities and for the Polish territory by the General Directorate of Environmental Protection (GDEP) (Stryjecki & Mielniczuk 2011). In early 2021, the EC issued an update to the 2011 guidelines on wind energy and Natura 2000 in accordance with the Action Plan for Nature, People and the economy (European Commission 2021). These studies are not legally binding; instead, they are intended to clarify and develop the existing legal provisions in this regard. The main purpose of the abovementioned publications was to popularize good environmental practices in the investment process, enabling to avoid and mitigate environmental as well as social conflicts, which in effect should significantly improve investment implementation processes. To date, these documents remain key to shaping the processes of environmental impact assessment – EIA (both strategic and investment) in the course of preparing investments in wind farms in Poland. An analysis of the abovementioned documents focusing on the environmental practice of onshore wind farm implementation is presented below.

The document published by the EC is more universal and does not go into such a high level of detail as the GDEP study, respecting the possible local variation of regulations, regarding environmental impact assessments, but falling within certain pan-European framework assumptions derived from the relevant directives. The study includes some procedures and good practices for mitigating, minimizing and compensating the impacts of wind investments, implemented both on land and (to a limited extent) in marine areas. At the same time, the study focuses in particular on mechanisms for assessing impacts on the objectives and objects of protection of the pan-European Natura 2000 network of protected areas, representing a liberal approach to the location of these investments within the network areas themselves and in their immediate vicinity. The document emphasizes the importance of individualized impact assessments on the environment and Natura 2000 sites, the result of which should determine the possibility or exclusion of a given investment. It should be noted that despite this pro-investment approach, a certain pan-European investment practice has been formed, which, as a rule, avoids (with the exception of a few EU countries) establishing these investments within and near areas of natural value. In addition, the impact of the implementation of planned projects on the permeability of ecological corridors and the coherence of the entire network is studied in detail in a separate procedure for assessing the impact on Natura 2000 areas even for the investments located at significant distances from the borders of areas covered by this protective status. This is a success of the document, which has been subjected to extensive consultation, including with industry organizations, and is a standard that is widely respected. The provisions of the EC's 2021 guidelines incorporate the latest advances in scientific knowledge and experience from as-built studies carried out across Europe, take into account the tremendous technological advances in the industry over the past decade and are supplemented by an expanded section on EIA for offshore wind farms.

The EC guidelines are supported by much more detailed regulations and guidelines of the member states. Similarly, this phenomenon also occurs in Poland, where several documents with the status of guidelines are currently in circulation and use, the history of implementation, timeliness and interdependence of which will be described below. GDEP guidelines relating to the same subject matter as the EC guidelines are currently in common circulation and use. This document should not be regarded as an alternative study, but as a complementary record of recommendations in the context of national legislation. The recommendations contained therein with a higher level of generality are supplemented by the context of Polish legislation, resulting from a number of legal acts (mainly the Environmental Protection Law, the EIA Act and the Nature Conservation Act). The GDEP guidelines, despite the initial objections of the wind energy industry regarding potential delays in the implementation of investments in the context of the new requirements, can be and are to this day a support for investors, while they were treated as a kind of guide by the officials of the regional directorates for environmental protection for many years, and despite the non-binding nature of their provisions were mostly fully enforced at the stage of preparing the EIA report and its subsequent supplements.

The GDEP study, compared to the EC guidelines, is more detailed and suggests certain practices that were not (and partially are still not) explicitly required by the provisions of Polish law. This gives additional material for the investor to consider, especially at the planning stage. It should be noted that there are also a number of guidelines from wind power industry and nature conservation organizations in operation (Polish Wind Energy Association 2008). In addition, the GDEP guidelines themselves were intended as a baseline study for guidelines on assessing the impact of wind farms on birds, bats and landscape, which were to be detailed in separate, dedicated guidelines published by GDEP, the first drafts of which were made public in 2011. These documents raised concerns not only in the investment industry but also in the expert communities of ornithology and chiropterology, as some of the recommended solutions were controversial as to their justification or even feasibility. At present there are the general guidelines of the GDEP for environmental impact assessments, published and official, the entire print run of which was rapidly sold out and the publication is now available only in digital form. The chiropterological and ornithological guidelines have retained the status of drafts (although their excerpts are used in procedures by regional directors of environmental protection on an ongoing basis). In this regard, it is worth pointing out that in 2017, GDEP published the recommendations for considering the impact of wind farms on the landscape in environmental impact assessment procedures.

The current status of the guidelines for wind power investment processes at various stages of their impact assessment is shown in Table 3.1.

TABLE 3.1

Overview of Current National Guidelines for Wind Power in the Context of Environmental Impact Assessments

Title	Year	Publisher	Status	Practical Implementation	Comments/ Need for Revision
Guidelines for assessing the impact of wind farms on birds	2011	GDEP	Draft made public	Partial	Need for revision
Guidelines for assessing the impact of wind farms on bats	2013	GDEP	Draft made public	Partial	Need for revision
Guidelines for forecasting environmental impacts of wind farms	2011	GDEP	Publicized, in use	Full	Need to adapt to the current legal status
Methodological proposal for the assessment of environmental determinants of the location of wind farms on a regional scale	2012	POLISH GEOGRAPHICAL REVIEW	Publicized, in use	Partial	Need to adapt to the current legal status
Recommendations for taking into account the impact of wind farms on the landscape in environmental impact assessment procedures	2017	GDEP	Publicized, in use	Partial	Review for practical implementation
Code of Good Practices	2019	PWEA	Publicized, in use	Limited by the 10H rule	Up-to-date

The 2011 GDEP guidelines represent the most synthetic study in the context of environmental impact assessments. However, in order to be fully applicable, they would need to be revised, particularly in terms of legislative changes made in the last decade. The issues worth considering when updating the aforementioned guidelines or preparing a new document based on a similar approach include:

- change in the selection of wind farm location – the current legal status excludes the implementation of investments based on the decision on development conditions, which was possible in 2011 – the so-called planning obligation;
- verification of the formal requirements for the project outline specification – the requirements of the guidelines went beyond the legal requirements at the time, but the scope of the required information in the project outline specification was in line with Article 62a of the EIA Act;
- verification of the procedure related to the issuance of a decision on environmental conditions, as the competent authority to issue the decision in question is the relevant regional director of environmental protection;
- updating recommendations on forecasting the range of acoustic impacts (in the audible and infrasound bands) with the practical experience of the past decade;
- updating basic information regarding the practices of pre- and post-construction monitoring;
- taking into account the changes resulting from current legislation, including the current state of the law on: protection of agricultural and forest land, planning procedures, obtaining construction permits, the RES Act, environmental impact assessment procedures, requirements under the Environmental Protection Law and its implementing acts and dependent standards, in particular the Distance Law.

However, it needs to be emphasized that the revision or update of existing guidelines has an informational and popularization value rather than a legislative one, as it does not affect the existence of certain legal bonds to the investor and the need to comply with existing laws, which must be implemented in the course of project implementation. The guidelines can contribute to better quality projects, speed up procedures and reduce the time or scope of required completions. This goal is to be achieved by promoting certain good practices. This was the premise behind the 2019 Code of Good Practices, prepared for local governments, the public and investors by a group of stakeholders in the form of the Union of Polish Banks, DNB Bank and local government organizations: Union of Rural Municipalities of Poland and Association of Renewable Energy Friendly Municipalities. This study, the most up-to-date of the existing ones on the national market, apart from general recommendations on planning and assessment procedures, also includes an elaboration on the issues normalized by law and sanctioned by investor practice. Among the most significant of these are public consultation, public participation and forms of securing title to land for investment. The approach to the investment by prioritizing its entire life cycle of the power plant (Life Cycle Assessment – LCA) (Matuszczak & Flizikowski 2015) is

also noteworthy; hence, the key issue of farm modernization for harmonious operation, as well as its demolition or repowering is also reflected in the content of this
document.

In addition, it should be noted that the ever-improving quality of projects, as a
result of both the learning effect of the Polish developer market, which is growing as
a function of time, and the increasingly stringent legal and guideline requirements,
will also be affected by the requirements of wind project financing institutions. With
the new auction support system coming into effect, investors are interested in minimizing the cost of the energy they generate in order to remain competitive in the auction. This means that these projects are often associated with an increasing financial
regime. In turn, obtaining favorable financing is currently equally dependent on project profitability as on project quality. In the market, the majority of projects are still
externally financed (usually through project finance), and the institutions providing
such support usually conduct assessment of project quality with due diligence, which
usually imposes very high standards on investments, both environmentally and in
terms of building public acceptance, as proper management of these issues significantly reduces project risk. There are a number of widely accepted international
standards in this regard, such as the Equator Principles; in addition, each financing
institution has its own corporate standards in this regard, often going well beyond the
literal legal requirements for investments. Thus, it can be concluded that in the near
future, when the price of energy from RES will be crucial, very high standards for
the implementation of wind farms in Poland will be primarily enforced by economic
and market realities (Table 3.2).

The Code of Good Practices (Polish Wind Energy Association 2019) is helpful in
the case of wind power development that promotes mutual understanding among all
stakeholders. In addition to a discussion of basic legal norms, it contains a wide range
of self-regulation mechanisms in the form of standards and rules of conduct from the
preparation of investments to the commissioning and long-term operation of wind
farms, which allow wind energy to become a fully socially friendly source of energy.
The code presents standards and rules of conduct, referring to well-established and
proven processes of wind energy development in the European Union, especially
those that have led to the formation of positive behavior in the process of conducting
investments in the development of wind farms and building good relations between
local governments, investors and local communities.

Chapter 3 can be summarized in the following points:

1. Wind power development in Poland is thoroughly and multilaterally regulated by universally applicable laws.
2. The key law for onshore wind energy "on investments in wind farms" functions together with other legal acts and the state's energy and environmental
   strategies and policies, in addition to national action plans for renewable
   energy as well as energy and climate. An element of the aforementioned
   law, which, after possible modification, may help in the further development of the renewable energy source in question, is reducing the required
   minimum distance of wind turbines from residential buildings.

TABLE 3.2

Comparison of the Environmental Guidelines in the Study Commissioned by EC in 2021 and GDEP in 2011, Taking into Account the Necessary Changes as Part of the Revision (Italics)

Criterion	2021 EC Guidelines	2011 GDEP Guidelines
Legal validity	They have no binding character.	They have no binding character.
	The guidelines are intended to approximate and interpret the provisions of the so-called Habitat and Bird Directives.	The guidelines relate to the interpretation of universally applicable laws.
	They are suggestions and development of the provisions of the abovementioned legal acts for good practice.	They are a suggestion aimed at rationalizing and improving decision-making processes.
	The document is an update of the EC guidelines of 2011.	
Areas covered by the provisions of the guidelines	Natura 2000 areas only.	All, including Natura 2000 areas.
Restrictions on the implementation of investment tasks	Natura 2000 areas do not exclude human economic activity; they only impose a formal framework that allows the preservation of biodiversity, which may be threatened by the effects of economic activity.	Resulting only from valid decisions on environmental conditions of the *competent regional director of environmental protection*, or a decision to refuse to determine the environmental conditions for the implementation of the project or to agree on the conditions for the implementation of the project in terms of its impact on the Natura 2000 area, or to decline them.
The need to protect species of prime importance to the community	Also outside Natura 2000 areas, especially in the case of potential wildlife corridor blockage.	To be considered for any investment.
	Deviations from the stated objectives of protection are possible only in the case of investments with high importance for public health or public interest.	

(Continued)

TABLE 3.2 (*Continued*)

Comparison of the Environmental Guidelines in the Study Commissioned by EC in 2021 and GDEP in 2011, Taking into Account the Necessary Changes as Part of the Revision (Italics)

Criterion	2021 EC Guidelines	2011 GDEP Guidelines
Assessment procedure	The relevant assessment may be coordinated with other environmental impact assessments, namely the environmental impact assessment for projects and the strategic environmental impact assessment for plans and programs, or may be incorporated into such assessments.	The assessment of the impact of a project on the object and objectives of protection of the Natura 2000 as part of the procedure for issuing a decision on environmental conditions (on the basis of the EIA report) or as a stand-alone document for the regional director of environmental protection being the basis for issuing an agreement on the conditions for implementation of projects in terms of impact on the Natura 2000 area.
The possibility of implementing a project having a significantly negative impact on the purpose and object of protection of the Natura 2000 area	Possible only under the condition of demonstrating the absence of technically feasible alternatives and with the demonstration of the realization of the overriding public interest (i.e., for human health, public safety, public service obligations, e.g., transportation and energy) and with optimal mitigation of negative impacts and thoughtful environmental compensation.	Possible provided that the prerequisites mentioned in Article 34 of the Law on Nature Protection are met, i.e., when the implementation of a project significantly affecting the objectives of protection of Natura 2000 area is supported by the requirements of overriding public interest, including requirements of social or economic nature, while there are no alternative solutions. It is necessary to implement natural compensation to ensure the coherence and proper functioning of the Natura 2000 network.
Ecophysiography	Not described in the guidelines.	The role of a well-executed ecophysiographic study as the basis for an environmental impact forecast indicated as part of the preparation of a draft local development plan or a study of land use conditions and directions.

(Continued)

TABLE 3.2 (Continued)

Comparison of the Environmental Guidelines in the Study Commissioned by EC in 2021 and GDEP in 2011, Taking into Account the Necessary Changes as Part of the Revision (Italics)

Criterion	2021 EC Guidelines	2011 GDEP Guidelines
Strategic planning	Strategic planning of wind power sector investments takes into account not only wind conditions but also technical feasibility of construction, connection to the power grid, distance of human settlements, landscape, nature conservation goals, etc. All of these conditions must be taken into account and can affect the feasibility as well as implementation of wind power projects. Planning is preceded by a strategic impact assessment, which also provides an appropriate framework for determining cumulative impacts.	The guidelines pay particular attention to the fact that under Polish conditions, before planning is started it is usually necessary to create a local spatial development plan (if the municipality does not have one) or amendments to the existing local spatial development plan or the study of spatial development conditions and directions, which usually do not provide for the possibility of wind energy development in the area to which they apply. *In the current legal order, there is no possibility of establishing wind farms based on a zoning decision.*
Environmental impact assessment stage of the project	The guidelines explicitly mandate that the environmental impact assessment of each project be carried out separately. It is pointed out that effective public participation in decision-making enables to express, and the competent authorities to take into account, the submitted proposals and opinions, thus increasing accountability and transparency in the decision-making process. Conducting a proper assessment involves the following steps: gathering information on the project and Natura 2000 sites; assessing the effects of project in light of the site's conservation objectives; determining whether the project may adversely affect the integrity of the site; considering mitigation measures (including monitoring).	According to the guidelines, all elements of the Wind Farm should be subject to environmental impact assessment, i.e., Wind farm, connection infrastructure (power and telecommunication cables), internal farm (transformer/switching station), access roads and maneuvering areas, construction facilities, assembly and storage yards (only during construction), optimally within one proceeding (if possible). If it is not possible to carry out a joint environmental impact assessment for all elements of the wind farm, a separate environmental impact assessment should be carried out for individual elements, taking into account their cumulative impact as a functional whole.

(Continued)

TABLE 3.2 (*Continued*)
Comparison of the Environmental Guidelines in the Study Commissioned by EC in 2021 and GDEP in 2011, Taking into Account the Necessary Changes as Part of the Revision (Italics)

Criterion	2021 EC Guidelines	2011 GDEP Guidelines

Anticipated impact on the natural and human environment	As indicated in the guidelines, there are many cases where well-designed and appropriately located investments are unlikely to have a significant impact, while in other cases multiple significant impacts are possible.
	The document lists the following factors that can negatively affect the environment within and around a wind farm:
	
	birds:
	
	1. habitat loss and degradation,
	2. disturbance and displacement,
	3. habitat fragmentation,
	4. collisions,
	5. barrier effect,
	6. indirect impacts;
	
	bats:
	
	1. loss and degradation of habitat and air corridors,
	2. disturbance and displacement,
	3. collisions,
	4. barrier effect,
	5. barotrauma,
	6. increased risk of collision due to night lighting,
	7. indirect impacts;
		Divided into construction, operation and decommissioning phases:
		
		1. construction and decommissioning stage:
		  – water pollution,
		  – air pollution,
		  – noise emission,
		  – electromagnetic field generation,
		  – soil pollution and waste generation,
		  – disruption of human living conditions,
		  – destruction of habitats and reduction of populations,
		  – visible changes in the landscape,
		  – impact on material assets, monuments, cultural landscape.
		
		2. operation stage
		I. Ornithofauna:
		1. risk of fatal collisions,
		2. loss and fragmentation of breeding and foraging habitats,
		3. creation of barrier effect.
		II. Chiropterofauna:
		1. destruction of winter or breeding quarters,
		2. crossing of flyways including migratory routes,

(Continued)

TABLE 3.2 (Continued)

Comparison of the Environmental Guidelines in the Study Commissioned by EC in 2021 and GDEP in 2011, Taking into Account the Necessary Changes as Part of the Revision (Italics)

Criterion	2021 EC Guidelines	2011 GDEP Guidelines
	valuable natural habitats:	3. loss of hunting grounds due to collisions (a phenomenon more pronounced in wooded areas) or bypassing the area.
	1. loss and degradation of habitats,	
	2. introduction of invasive alien species during the construction phase,	III. Acoustic environment
		IV. Infrasound
	3. creation of habitats away from the wind farm in order to attract birds to these habitats and draw them away from the wind farm,	V. Electromagnetic field and radiation
		VI. Landscape
	4. creation of habitats on intensively used agricultural land by providing less intensively used remaining areas,	1. zone I (up to 2 km from the WF)
		2. zone II (1–4.5 km from WF)
	5. changes in microclimate,	3. zone III (2–8 km from WF)
	6. soil compaction,	4. zone IV (more than 7 km from WF)
	7. indirect impacts.	VII. Value of the property[1]
Requirements for monitoring	The best model for environmental monitoring is to measure impacts before and after they occur. In the area likely to be affected by the project and at control sites not affected by the project, baseline data should be collected (before the start of the project) using a standard method. Then, during operation of the project, data is collected in the project area and at control sites using the same method. Like the collection of output data, monitoring should be designed using a standard approach to data collection and statistical analysis.	Preliminary monitoring of potentially sensitive natural elements and possible social conflicts is recommended already at the screening stage.
		3. *For birds – use of the Guidelines for assessing the impact of wind farms on birds (GDEP, 2011);*
		4. *For bats – use of the draft Guidelines for assessing the impact of wind farms on bats (GDEP, 2013).*
		It should be noted that these are recommendations of good practice, but not functioning as binding law, so other monitoring methods justified by current knowledge are also acceptable.
		Annual pre- and post-investment monitoring is recommended where warranted.

(Continued)

TABLE 3.2 (Continued)

Comparison of the Environmental Guidelines in the Study Commissioned by EC in 2021 and GDEP in 2011, Taking into Account the Necessary Changes as Part of the Revision (Italics)

Criterion	2021 EC Guidelines	2011 GDEP Guidelines
Procedures for the implementation of wind power investments	The guidelines refer to the procedure, which includes three main stages: 1. **stage one: preliminary screening.** First, it must be determined whether the plan or project is directly related to or necessary for the management of the Natura 2000 site in question, and second, if not, whether it is likely to have a significant impact on the site; 2. **stage two: appropriate assessment.** Appropriate assessment of the effects on the site from the perspective of the site's conservation objectives is conducted. 3. **stage three: derogation from Article 6(3) of the Habitats Directive.** The third stage applies when, despite a negative assessment, it is not proposed to reject the plan or project, but to further evaluate it. In this case, derogations are allowed under certain conditions, which include the demonstrated lack of alternatives and the existence of an overriding public interest in favor of the project. This requires the adoption of appropriate compensatory measures to ensure the overall coherence of the Natura 2000 network.	**Procedures to obtain a decision on environmental conditions** **Type I investments:** 1. qualification of the investment for Group I, 2. preparation of an environmental report, 3. submission of an application for a decision on environmental conditions, 4. initiation of proceedings by the administrative body, 5. application of the authority for an opinion/ determination on the conditions for the implementation of wind farms, 6. issuance of opinions and conduct of arrangements by cooperating authorities, 7. conduct of the public participation procedure by the leading authority, 8. issuance of a decision on environmental conditions, 9. making the decision on environmental conditions public.

(Continued)

TABLE 3.2 (*Continued*)

Comparison of the Environmental Guidelines in the Study Commissioned by EC in 2021 and GDEP in 2011, Taking into Account the Necessary Changes as Part of the Revision (Italics)

Criterion	2021 EC Guidelines	2011 GDEP Guidelines
		Type II investments:
		1. qualification of the investment for Group II,
		2. preparation of a project information sheet,
		3. submission of an application for a decision on environmental conditions,
		4. initiation of proceedings by the administrative body,
		5. requesting an opinion from the authority on the necessity of conducting an environmental impact assessment and determining the scope of the EIA report,
		6. issuance of opinions by the interacting authorities,
		7. issuance of an opinion by the leading authority on the necessity of conducting an environmental impact assessment and determining the scope of the EIA report, or a decision on the lack of necessity of conducting an environmental impact assessment,
		8A. issuance of a decision on environmental conditions (if a decision on the lack of need for an environmental impact assessment has been issued),
		8B. Preparation of an EIA report and submission of it to the leading authority (if a decision on the need for an EIA has been issued),

(Continued)

TABLE 3.2 (Continued)

Comparison of the Environmental Guidelines in the Study Commissioned by EC in 2021 and GDEP in 2011, Taking into Account the Necessary Changes as Part of the Revision (Italics)

Criterion	2021 EC Guidelines	2011 GDEP Guidelines
		9. requesting the opinion/agreement of interacting bodies from the leading authority,
		10. issuance of opinions and agreements of the interacting bodies,
		11. conduct of public participation procedure by the leading authority,
		12. issuance of a decision on environmental conditions,
		13. publicizing the decision on environmental conditions.
		Procedure for re-evaluation of environmental impact (carried out, inter alia, before obtaining a construction permit, approval of a construction project, decision on permission to resume construction work)
		1. Determination of the need for re-evaluation of environmental impact (if such a need is identified in the decision on environmental conditions after the assessment of the impact of the project on the environment, at the request of the interested party or in the decision of the authority that issues the construction permit, usually required if the construction project deviates from the design presented in the decision on environmental conditions).

(Continued)

TABLE 3.2 (*Continued*)

Comparison of the Environmental Guidelines in the Study Commissioned by EC in 2021 and GDEP in 2011, Taking into Account the Necessary Changes as Part of the Revision (Italics)

Criterion	2021 EC Guidelines	2011 GDEP Guidelines
		2. The EIA report for reassessment of environmental impact is more detailed (it takes into account detailed data arising from the construction project, determines the degree and manner of taking into account the requirements for the environment contained in the decision on environmental conditions and other opinions/decisions issued for the project).
		3. The EIA report is submitted to the authority issuing construction decisions.
		4. The decision-issuing authority, upon receipt of the EIA report, requests the regional director of environmental protection to agree on the conditions for implementation of the project.
		5. The regional director of Environmental Protection applies to the State Sanitary Inspection for an opinion.
		6. The authority issuing the construction decision conducts public consultations at the request of the regional director of environmental protection.
		7. The conclusions of the consultations are forwarded by the authority issuing the construction permit to the regional director of Environmental Protection, who, on this basis, issues a decision to agree on the conditions for implementation of the project.

(*Continued*)

TABLE 3.2 (Continued)
Comparison of the Environmental Guidelines in the Study Commissioned by EC in 2021 and GDEP in 2011, Taking into Account the Necessary Changes as Part of the Revision (Italics)

Criterion	2021 EC Guidelines	2011 GDEP Guidelines
		Investments in Natura 2000 areas – are carried out under the standard procedure for obtaining a decision on environmental conditions with an EIA report enriched with the determination of the impact of the project on Natura 2000 areas. For the investments that do not belong to the I and II group of projects from the EIA Regulation, the procedure is as follows: 1. The administrative authority decides whether the wind farm may have a potentially significant impact on the Natura 2000 area. 2. If the potential significant impact on the Natura 2000 area is determined, the administrative body shall require the investor to submit the following documentation to the competent regional director of Environmental Protection: • application for a decision, • project information sheet, • a copy of the cadastral map, certified by the relevant authority, covering the area of construction implementation and the area of potential impact. 3. On the basis of the documentation presented, the local regional director of Environmental Protection issues a decision on the need, or lack thereof, to carry out an assessment of the impact of the project on the Natura 2000 area. *(Continued)*

TABLE 3.2 (Continued)

Comparison of the Environmental Guidelines in the Study Commissioned by EC in 2021 and GDEP in 2011, Taking into Account the Necessary Changes as Part of the Revision (Italics)

Criterion	2021 EC Guidelines	2011 GDEP Guidelines
		4. In a separate document, the regional director of environmental protection imposes the obligation to submit two copies (hard copy and digital recording) of the report on the impact on the Natura 2000 area, which should be limited to the determination of the impact of the project on the Natura 2000 area.
		5. The public participation procedure is carried out by the administrative authority at the request of the regional director of Environmental Protection.
		6. The administrative body forwards the collected comments to the reconciliation authority.
		7. The regional director of Environmental Protection agrees on the conditions for implementation of the project, in the justification he/she is obliged to present the impact of the results of the consultation and the submitted report on the issued conditions.
		8. If the assessment of the impact of the project on the Natura 2000 area shows that the investment will have a significant negative impact on this area and there are no prerequisites of overriding public interest, the regional director of Environmental Protection refuses to agree on the conditions for the implementation of the project

(Continued)

TABLE 3.2 (Continued)
Comparison of the Environmental Guidelines in the Study Commissioned by EC in 2021 and GDEP in 2011, Taking into Account the Necessary Changes as Part of the Revision (Italics)

Criterion	2021 EC Guidelines	2011 GDEP Guidelines
Opportunities to reduce and/or prevent annoyance from wind farms	1. Planning – setting and establishing wind turbines; 2. Infrastructure design: number of turbines and technical specifications (including lighting); 3. Scheduling: avoiding, limiting or dividing construction activities into stages during environmentally sensitive periods; 4. Limiting turbine operation and switching wind speeds: defining turbine operation times; 5. Using deterrents.	Examples of opportunities to avoid or reduce the annoyance of wind farms: 1. performing a specialized analysis of acoustic impacts and electromagnetic field and radiation at the design stage; 2. drawing up ornithological, chiropterological and habitat inventory at the design stage; 3. conducting multi-criteria analysis of possible alternatives and selection of the most favorable one; 4. ensuring adequate construction supervision at the investment stage; 5. adopting a well-thought-out method of storing, neutralizing and transporting waste products generated at the construction stage; 6. ensuring the free passage of birds and bats through proper placement of towers; 7. painting the structures with light-colored matte paints, which reduces glare and increases their visibility; 8. carrying out construction work outside the period of breeding and migration of birds while logging additionally outside the growing season.

(Continued)

TABLE 3.2 (*Continued*)
Comparison of the Environmental Guidelines in the Study Commissioned by EC in 2021 and GDEP in 2011, Taking into Account the Necessary Changes as Part of the Revision (Italics)

Criterion	2021 EC Guidelines	2011 GDEP Guidelines
		In addition, consideration should be given to:
		1. use of safe substances,
		2. energy efficiency criterion,
		3. rational use of fuels and raw materials,
		4. limitation of production and recovery of by-products and waste products,
		5. type, extent and volume of emissions,
		6. use of favorable technical solutions adapted from industry, scientific and technological progress.
Nature compensation methods	Resulting from provisions:	In each case, they should take into account the specific impact of a particular investment.
	1. restoration or expansion of habitat within an existing Natura 2000 site,	
	2. restoration of a habitat on a completely new site or by expanding the area of an existing habitat for inclusion in the Natura 2000 network,	
	3. designation of a new Natura 2000 site (of the same type, to preserve the coherence of the network).	

1. According to Article 66 of the EIA Act, which defines the content of the EIA report, one cannot derive the necessity to examine the impact of the project on the value of real estate (judgment of the Regional Administrative Court in Gliwice of March 18, 2015, ref. no.: II SA/Gl 1530/14).

3. It is recommended, based on the positive results of the guidelines used so far for the preparation of EIA reports for wind farms, to modernize these guidelines and refine them (in particular drawing on the experience of as-built monitoring carried out on Polish wind farms over a 10-year period) and (after revision) publish the draft guidelines in other areas (e.g., concerning ornithofauna and chiropterofauna).

4. Industry guidelines developed in the course of extensive consultations with all stakeholders in the process are widely respected and used; hence, it is recommended that there be extensive industry consultation of any new studies in this regard.

5. It is also advisable to make fuller use of ecophysiographic studies, not only for the preparation of reports and forecasts of the EIA but also for the purposes of spatial planning in the municipality.

6. It is recommended to place greater emphasis than before on consultations with the public during the implementation of investments in wind energy.

4 Mitigating Negative Impacts and Ensuring Safety of Residents Living in the Vicinity of Wind Farms

4.1 MINIMUM DISTANCE OF WIND TURBINE(S) FROM RESIDENTIAL BUILDINGS

Certainly, wind turbines emit noise during their operation, which decreases with distance from the source. A low health risk will occur when noise limits are not exceeded. In contrast, no risk will occur when noise emission levels are below human auditory and vibrational perception. Assessment of the impact of any noise source is carried out with the aim of minimizing the risk of loss of health or damage to the environment. Complete risk elimination is possible only by not undertaking the investment. If there is no danger, there is no risk. Therefore, if the investment is to be undertaken, then the minimum distance of the wind turbine from residential buildings that maintains acceptable limits of noise in the environment must be determined. This does not mean that the sounds emitted by the turbines will be inaudible, but it means that they will be at a level that ensures acoustic comfort and low risk of harmful impact. To this end, procedures are needed to determine permissible levels, emission forecasting methods and measurement methods of control.

As shown in Section 2.1, the permissible values of environmental noise from wind turbines, included in the group "Other facilities and activities that are a source of noise" in the regulation (Regulation – noise 2007) should be considered adequate.

In the numerical calculations carried out at the design stage of the farm, the most unfavorable case of emissions is assumed. The maximum sound power level of the turbines used during the entire period of their operation is assumed for the calculations. It is assumed that the acoustic power levels of the turbines adopted for the calculations will not be exceeded in practice.

The purpose of the controlled noise measurements is the assessment of the operating wind farm in terms of risk to the environment and people. Field measurements are a verification of the results of numerical calculations and acoustic assessment of the investment task, determining the location of individual wind turbines.

In view of the decrease in the recorded sound level with the distance from the source (wind turbine), it is important to determine the safe distance of the wind

DOI: 10.1201/9781003456711-5

turbine from residential buildings. According to the authors, this distance should always result from simulations of noise propagation and be verified during control measurements in the environment. If, as indicated in Section 2.1.3, a safety buffer (3 dB) is accounted for at the forecasting stage, which may be used following post-implementation noise measurements, the risk of exceeding environmental noise limits will be negligible. Section 1.1.9 (Table 1.7) summarizes the results of post-implementation measurements conducted around wind farms in Poland. At a distance of 500 m, the permissible noise level was exceeded only once, out of 28 wind farms surveyed by the authors of the study (approximately 100 measurement sessions were carried out). Due to noise-reducing operations performed (changes in the mode of operation of the device "mods"), even this farm no longer exceeds noise limits in the environment.

The fact that the noise heard at a distance of 500 m from the wind turbine does not exceed the permissible levels of noise in the environment (as specified in the Regulation of the Minister of Environment (Regulation – noise 2007)) is evidenced, i.e., by the results of computer simulations using CadnaA software, conducted by Guarnaccia et al. (2011).

Figure 4.1 shows that already at a distance of about 85 m from the wind turbine, the noise level is close to 50 dB, which corresponds to the standards for acceptable noise levels in single-family residential areas during daytime hours. In contrast, other studies show that at a distance of 500 m from a wind turbine it is below 40 dB. Such noise levels should not cause negative health effects, including for sensitive people (Guarnaccia et al. 2011). Others indicate that at a distance of 500 m the noise level does not even exceed 35 dB (Knopper & Ollson 2011).

In some countries, despite the noise propagation simulations carried out at the stage of designing the location of a wind farm, the minimum distances of a wind

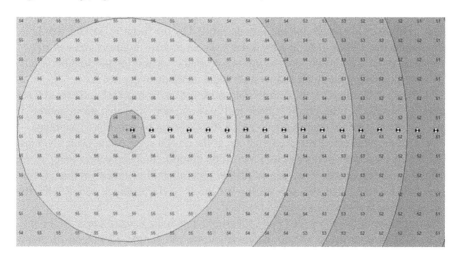

FIGURE 4.1   CadnaA audible noise map (values are given in dB) for a point source of noise with LW = 103.5 dB (wind speed = 8 m/s), height = 70 m, absorbing ground and high directionality. The grid is evaluated at a height of 2 m from the ground, the distance between the two receivers (black and gray circles) is 5 m, and their height is 4 m.

turbine from residential buildings are mandatory. In some countries these values are fixed, in others they are negotiable at the level of local law. Therefore, the summary presented in Section 2.1.2 (Table 2.1) illustrates only the preferences of state governments in terms of environmental protection around wind turbines. The purpose of such a solution is to indicate the danger and the desire to minimize the risk of impact. The proposed distance is similar to an orange traffic light, indicating the need for caution. The minimum distance must not be too large so as not to block the possibility of investment, unless the purpose of its introduction is not to minimize the risk of impact.

With the above in mind, and taking into account the results of field measurements of noise around wind farms carried out by the authors of the study, the warning distance may be set at 500 m. However, the final distance should always be determined by calculation methods and verified by field measurements.

In order to determine the minimum distance for establishing a wind farm away from residential buildings, all the previously described impacts of such a farm on the surrounding immediate environment should be considered together. The following "risk map" of the potential impact of a wind farm on the environment shows the change in the probability of occurrence of various impacts depending on the distance of the location of the nearest EW from the observation point. Individual risk symbols (H, I, M, E, D, C, L, A) appear when the estimated probability value exceeds 0.1%. The probabilities were estimated based on authors' own observations and calculations, as well as publicly available information (Figure 4.2).

In summary: the impact of acoustic noise has the main effect on the determination of the allowed minimum distance of locating a wind farm in relation to residential development. The impact of other effects ceases to be a nuisance at a distance of 500 m.

Probability of impact [%]	up to 100	up to 200	up to 300	up to 400	up to 500	up to 600	up to 700	up to 800	up to 900	up to 1000
Very high 80–100	H									
High 61–80		H								
Moderate 41–60			H							
Low 21–40				H						
Very low 0–20	I, M, E, D, L	I, M, L	I, M, L	I, M, L	H, I, M L	H, I, M	H, I, M	H, I, M	H, I, M	H, I, M

Distance from wind farm [m]

Legend: **H** – audible noise, **I** – infrasound, **M** – light flicker, **E** – electromagnetic field, **D** – vibration, **C** – detachment of turbine parts, **L** – ice throw, **A** – malfunction (fire, disintegration)

FIGURE 4.2   "Risk map" of potential environmental impact.

4.2 SUGGESTIONS FOR MODIFYING LAWS AND GUIDELINES

In mature democracies, the evolving economy has set new paths for development and the law has kept pace with these changes and prepared the foreground for further development by listening to what entrepreneurs expect.

The Polish economy needs renewable energy sources, and the wind energy sector needs stable and predictable operating conditions. Such a system of legal acts, guidelines and codes of good practice should not only cover all areas that are related to the construction and operation of wind farms, including environmental issues, extremely important consultations with the local community but also outline the formal obligations of investors and their commitments to other stakeholders.

Currently, the wind energy industry needs modifications to some legal acts, including:

- The Act of May 20, 2016, on investments in wind farms (Journal of Laws 2016, item 961), in the part concerning the determination of the minimum permitted distance of wind turbines from human settlements. The justification is the entire content of this monograph.
- Regulations of the Minister of Climate and Environment dated September 7, 2021, on the requirements for conducting measurements of emissions (Journal of Laws 2021, item 1710) (Regulation – emissions 2021). It is suggested to add Annex 9, dedicated to measurement methodologies used around wind farms. Guidelines in this regard are given in Section 2.1.4.
- Guidelines for forecasting environmental impacts of wind farms (Stryjecki, Mielniczuk, GDEP 2011). It is necessary to set the guidelines in the context of the most up-to-date versions of the detailed guidelines, as the current ones refer to the documents that were best practice at the time of their publication, but have been superseded by others or have naturally become obsolete. Recommendations for predicting the extent of acoustic impacts (in the audible and infrasound bands) should be updated with the practical suggestions made in Sections 2.1.3 and 2.1.5, as well as include basic information on pre- and post-construction monitoring practices (Section 2.1.4).
- The Act of December 21, 2000, on technical supervision of the operation of wind farms.
- Polish Standard PN-B-02151-2:2018-01, Building acoustics "Protection against noise of rooms in buildings. Permissible values of sound levels in rooms." In order to implement the recommendations contained in Section 2.1.5, it is necessary to determine the permissible levels for low-frequency and infrasound noise.

4.3 GUIDELINES AND RECOMMENDATIONS

In order to present guidelines and recommendations in a clear and complete manner, it is necessary to familiarize oneself with the current state of the law and other materials that may be helpful in this regard. While directives, laws, regulations, BAT conclusions are available, supplementary materials require laborious collection, familiarization and thorough evaluation. In the latter case, it is necessary to conduct a thorough analysis of domestic and foreign documents (mainly from the EU area) so far issued and prepared for publication guides and codes of good practice, both by state and local administrations at various levels and by non-governmental organizations (chambers and associations, including industry associations), discussing various aspects of the environmental impact assessment report, the decision on the environmental conditions of the project, with particular emphasis on dialogue with the local public.

4.3.1 Development of the Report on the Environmental Impact of Wind Farms

In order to make the execution of the EIA report more efficient and faster, among other things, it is necessary:

- precise elaboration of the scope of individual requirements for EIA reports by the Law of October 3, 2008, "on the provision of information on the environment and its protection, public participation in environmental protection and environmental impact assessments" and mandatory compliance with it throughout the country. In particular, this applies to ecophysiographic studies, descriptions of natural elements, natural inventories, measurement methodologies, permitted interpretations, forecasting methods, permitted methods and scopes of natural compensation, analyses of potential social conflicts. These requirements are mentioned in the law and regulations, but are applied and interpreted differently by both the competent authorities and investors, in different parts of Poland,
- preparation of ecophysiographic studies, in cooperation with the authority issuing the environmental approval, in such a way that the content of this study is useful for the creation or modification of local spatial development plant or study of conditions and directions of spatial development (synergy of the interests of the municipality and the investor),
- facilitating the access of the investor and the public party to the information resources of the authorities competent to issue the EIA and other institutions, at the stage of preparing the EIA report,
- striving to ensure that a single EIA report (and then also single environmental approval) covers all elements of the wind farm infrastructure, i.e., wind farms, connection infrastructure (power and telecommunications cables), internal farm transformer/switching station, access roads and maneuvering areas, construction facilities, assembly and storage yards (only during construction),
- in justified cases (e.g., a single wind farm located in a non-collision), the project outline specification, rather than an EIA report, should be sufficient,
- close cooperation of the investor with the authority competent to issue the EIA throughout the wind farm construction cycle (especially in the initial phase) and the local community,

- in terms of acoustics, the scope of the report must include an analysis of the impact of infrasound and acoustic noise. The method of analysis should be in accordance with the recommendations found in Sections 2.1.3 and 2.1.5. Regardless of the detail of the report, the decision on monitoring should take into account the suggestions found in Sections 2.1.4 and 2.1.5,
- other guidelines for more efficient and quicker preparation of the EIA report are included in Chapters 2 and 3 of this study.

4.3.2 Assessment of the Report on the Environmental Impact of a Wind Farm (Forecast)

Regarding noise emissions, the guidelines for forecasting environmental impacts of wind farms should be completed.

The analysis in the report for environmental impact assessment should meet the following assumptions:

- Adoption of permissible noise levels from the group "Other facilities and activities that are a source of noise" in accordance with the regulation (Regulation – noise 2007) (Journal of Laws 2014 item 112, Table 1).
- Evaluation of the results of simulations conducted according to ISO 9613-2 on the basis of:
  - spectral (one-third octave or octave) distribution of sound power levels,
  - application of the method of general evaluation of ground influence, ground index $G = 0.7$.
- The designer should demonstrate in the acoustic analysis that after the construction of the wind farm there is the possibility of reducing the acoustic power of the turbine or applying a technical solution on the turbine that reduces noise emissions by 3 dB.

The analysis for the local spatial development plan of the LSDP should meet the following assumptions:

- If the plan area includes only agricultural land for the wind farm:
  - Evaluation of the results of simulations conducted according to ISO 9613-2 on the basis of:
    - single-number sound power level,
    - application of the "alternative" method of assessing the impact of the land (without the G index), which determines slightly overestimated noise extents that provide an opportunity for optimization in the future,
  - An indicator implemented into the local spatial development plan should be the impassable building lines, both for wind turbines and residential development.
- If the plan area includes areas with different urban functions:

- the indicator input into the local spatial development plan should be the outline of the wind farm located 500 m away from residential areas, which is the impassable building line for the turbines,
- explicit prohibition of residential development in the buffer zone of 500 m from the wind farm outline.

4.3.3 Monitoring the Environmental Impact of a Wind Farm

In terms of acoustic impact, a uniform methodology for taking measurements is important. It is possible to temporarily apply the recommendations presented in Section 2, but ultimately there should be Annex 9 in the Regulation of the Minister of Climate and Environment of September 7, 2021, on the requirements for conducting measurements of emissions (Journal of Laws 2021 item 1710) (Regulation – emissions 2021), describing in detail the methods for measuring noise around a wind farm. The methodology should take into account the guidelines contained in Section 2.1.4.

The assessment of infrasound noise remains an open question. It should be considered whether, given the relatively low levels of infrasound generated by wind turbines, often below the threshold of perception, there is a need to study it in the environment. Due to the numerous interferences from infrasound naturally generated in the environment, conducting field assessments seems to be a difficult task to accomplish. It is possible, following the Danish model, to perform infrasound noise assessments in buildings (*see proposals in* Section 2.1.5). As of today, the results of such measurements do not indicate a significant health risk caused by infrasound. Further research in this area is recommended.

In turn, the issue of measuring low-frequency noise, which has not been described in any aspect in Poland at present, needs to be regulated. Sounds in frequencies above 40 Hz recorded in the vicinity of wind farms have levels higher than the thresholds of human auditory perception and can be perceived by humans. The frequency range, the acceptable levels in the adopted range and the method of measurement should be described, which takes time. Until the company establishes its own methodology for assessing low-frequency noise, the proposals in Section 2.1.5, based on Danish solutions, are recommended and they are relatively easy to implement.

4.3.4 Post-implementation Analysis and a Detailed Method of Preparing such an Analysis for Wind Farms

The recommendations outlined in Section 2.1.4 can be applied on an interim basis, but ultimately there should be a detailed description of the method for measuring noise around the wind farm.

4.4 TECHNOLOGIES AND OTHER PREVENTIVE MEASURES

As already mentioned several times in this monograph, the currently manufactured wind turbines have the ability to reduce the level of acoustic power, i.e., the ability to reduce the noise produced during their operation.

Noise reduction of turbines is realized by:

a. Changes in blade angle, known as "Mods" or "Adjustments," which make it possible to reduce sound power levels by up to several decibels. Each turbine has details of sound power level reduction for subsequent "Mods" written in its technical documentation. These are very common methods of reducing noise during turbine operation, which are implemented through software and which are switched on automatically in situations of defined wind speeds at nacelle height. These are maintenance-free systems.

b. Appropriately designed serrations installed on the trailing edges of the blades, which minimize the generation of aerodynamic noise. The effectiveness of the serrations is currently about 2 dB.

c. Repowering of wind turbines, which in acoustic terms means replacing existing turbines with new equipment, which are fewer in number and may be quieter. The greatest advantage of repowering is to reduce the number of turbines within the operation of an existing wind farm after a certain period of operation of the investment – for example, after 20 years. This is when some of the existing turbines are dismantled and replaced with more modern equipment. Obviously, the new turbines will be more efficient, which will translate directly into fewer turbines and thus reduced noise ranges near the nearest residential buildings.

The recommendations presented in the monograph are intended to improve and standardize the environmental impact assessment of wind turbines, at the stage of forecasting and monitoring. It is important to be aware that there will be new types of wind turbines, the impact of which is unknown today. The trend in technological development of turbines is toward increasing productivity and reducing impact (especially noise). However, the impact of wind turbines needs to be monitored and necessary adjustments made on an ongoing basis as to their evaluation methods. It seems reasonable for the National Centre for Research and Development (NCRD) to support annually the research projects aimed at improving the calculation and measurement procedures used in the evaluation of wind farms, at solving various problems related to the operation of wind farms (such as those signaled in this chapter) and also at reducing their impact on the environment and especially on human health.

4.5 SUSTAINABILITY OF WIND POWER – PROCESSING END-OF-LIFE WIND TURBINE BLADE COMPOSITES

The wind power industry, like other industries, must adapt to the general trend of opposing climate change, including decarbonization. This takes on particular importance in light of efforts to significantly scale up wind projects and related industrial activities. Moreover, governments in many countries are taking steps to decarbonize industrial value chains, from green energy procurement to implementing zero-emission building standards. In short, it can be said that:

> The wind turbine industry needs to optimize technology and processes, improving existing solutions and introducing new ones, minimizing waste and decarbonizing the supply chain, which includes materials from "difficult to decarbonize", yet carbon-intensive sectors, such as steel and cement production.

A full life cycle analysis of the substrates and products consumed and produced in wind power and related industries includes emissions to air, water and land from a wind project at all stages of production, transportation, installation and decommissioning. The aforementioned analysis shows that the carbon payback period for wind power is much shorter than for coal-fired farms. According to 2016 data, it is about 5.4 months for a 2 MW onshore turbine and about 7.8 months for a 6 MW offshore turbine. In this respect, wind power surpasses hydroelectric and solar power (Global Wind Report 2021). The quoted report estimates that the production and installation stages are responsible for more than 90% of the total carbon dioxide emissions associated with an onshore wind farm and for 70% of the carbon dioxide emissions for an offshore wind farm (with the additional contribution of shipping).

It can be assumed that 80%–85% of the total weight of a wind turbine consists of recyclable materials such as steel, iron, copper and aluminum. Methods for their use are generally known and their application does not pose major difficulties. However, the remainder is made up of materials that are difficult to recycle commercially, such as carbon fiber or fiberglass composites, plastics and resin, mainly building rotor blades, which are estimated to have a lifespan of up to 25–30 years.

Composites can be broadly defined as a set of two or more materials that, when combined, have better properties than they do on their own. They usually consist of a reinforcing material (glass, carbon and other fibers) surrounded by a matrix material (resins). The fibers can take on several different weave architectures, resulting in layers that can be stacked and placed in different orientations to give the resulting composite product optimal performance for the chosen application. There are two main forms of such resins: thermosets, which form an irreversible solid polymer and are most commonly used for wind turbine blades, and thermoplastics, which can be melted and recycled. Fiber-reinforced composites are used in a wide range of applications, such as vehicle components, doors, bathtubs and wind turbine blades, among others.

This general characterization can be helpful in understanding the technologies used in the management of composite materials, both at the stage of product creation and its processing when it loses its original parameters.

The waste management hierarchy for wind blades can be presented as follows:

- Prevention: extending the life of the design or wind blade,
- Reuse: its sale on the used blade market,
- Repurpose: remanufacturing for use in new products,
- Recycling (*recycling*): shredding, grinding and milling to a filler fraction for FRP (*Fiber-Reinforced Plastic*) or concrete,

- Material recovery: pyrolysis, thermolysis, solvolysis to recover polymer resins, fibers or gases for energy use,
- Co-processing in cement kilns: substitution of natural raw materials,
- Incineration: with or without energy recovery associated with ash deposition in a landfill,
- Storage.

The wind power industry, supported by various governing bodies of the European Union and individual countries, has initiated several research and implementation programs aimed, among others, at producing rotor blades with materials that are more easily recycled in a cost-effective manner and at increasing the use of the recycled, decommissioned blades. The aforementioned initiatives have also been followed by similar programs at the individual company level. At present, the following initiatives, among others, are being implemented in Europe at various stages of development:

DecomBlades (Decom Blades) (recycling of rotor blades)

DecomBlades seeks to establish the basis for the commercialization of recycling technology, so the partners (Denmark's leading wind turbine manufacturers, including, among others, Vestas, Oersted, LM Wind Power) participating in this project have focused their efforts on three paths that are the most technologically mature, as well as the most cost-effective (ETIP 2019):

- pretreatment of the material (cutting by means of, i.e., water jet, diamond wire and hydraulics, as well as preparation for transportation) and its mechanical grinding,
- co-processing in cement production (in calciners and rotary cement kilns – as an alternative fuel component and as a clinker component),
- pyrolysis – during thermal treatment of composites under anaerobic conditions, resins are converted into liquid (e.g., used as fuel or for further use in the chemical industry) and gaseous phases (e.g., used for heating or electricity generation), whereas the solid phase corresponds to fibers (glass and carbon for reuse) and ash.

ZEBRA (*Zero wastE Blade ReseArch*) (ZEBRA 2019)

The project is led by the French research center IRT Jules Verne, which has managed to bring together industrial companies and technical centers (Arkema, Canoe, Engie, LM Wind Power, Owens Corning, Suez), representing the full value chain: from materials development, to blade production, to wind turbine operation and decommissioning, and finally recycling of decommissioned blade material.

LM Wind Power is designing and manufacturing two prototype blades using Arkema's Elium resin to test and verify the behavior of the composite material and its suitability for industrial production. Simultaneously, other project partners will focus on developing and optimizing the manufacturing process using automation to reduce energy consumption and production waste. ZEBRA partners will then investigate methods for recycling the materials used in the prototype blades. Life cycle analysis will assess the environmental and economic viability of continuing to use the thermoplastic material used in the production of blades for future wind turbines.

CETEC (*Circular Economy for Thermosets Epoxy Composites*) (CETEC 2021)

A coalition of established research institutions and industrial players (Vestas Wind Systems A/S, Olin Corporation, Danish Technological Institute and Aarhus University) partially funded by Innovation Fund Denmark (IFD) has developed a two-stage technology in which:

- in the first stage, thermoset composites are decomposed into fibers and epoxy resins;
- in the second stage, through a novel chemocycling process, the epoxy resin is decomposed into base components, which can then be reused in the manufacture of new wind turbine blades.

Circular Wind Hub (need to establish new closed-loop rules to increase recycling of wind turbine components) (Circular Wind Hub 2019)

Moonshot Circular Wind Farms (seeking solutions to secure wind farms in materials critical to their development) (Moonshot Circular Wind Farms 2020).

Both projects are supported by the Dutch Ministry of Economic Affairs and Climate Policy. They are motivated by the processing of materials from decommissioned wind farms. To do this with the highest possible quality, it is necessary to:

- establish new closed-loop economy policies to improve the life cycle of wind turbines,
- develop knowledge for optimal end-of-life strategies,
- develop industrial infrastructure capable of handling increasing numbers of wind turbines.

DecomTools (decommissioning/repowering offshore wind farms) (DecomTools 2020)

This is an initiative of scientific and research institutions, commercial companies, technical infrastructure units, pro-environmental NGOs and local government units from the North Sea basin. While the decommissioning/repowering processes of onshore wind farms are already reasonably well understood, the end-of-life experience of offshore wind parks is limited. This project bridges this gap by developing appropriate software and developing eco-innovative concepts that:

- reduce decommissioning costs by 20% and environmental footprint costs by 25% (measured in CO_2 equivalent);
- increase the know-how and expertise of the North Sea stakeholders involved.

Tests will be verified by pilot models and working tools from the areas of logistics, safety, ship design and recycling. The combination of innovative and already available technologies should address some of the main aspects of the challenges of decommissioning, including optimization, of existing maritime/land (port) infrastructure. Transnational cooperation and multidisciplinary competence of partners, representing many sectors of life and the economy, should improve innovation and technology transfer in this specific niche area and help the wind power industry achieve a higher degree of sustainability.

Dreamwind (developing new sustainable materials that can be (re)used in the manufacture of wind turbine blades; project already completed) (Dreamwind 2020)

Dreamwind (Designing REcyclable Advanced Materials for WIND energy) is a collaborative research project between the University of Aarhus, the Danish Institute of Technology and Vestas Wind Systems to develop new sustainable materials that can be used in wind turbine blades. The project is partly funded by the Danish Innovation Fund.

The development of new materials is viewed from the perspective of a closed-loop economy. Looking at materials development in a broader context focuses on developing a product that is economically and resource sustainable. Development is based, in part, on producing materials that can be disassembled after use and, in part, on incorporating organic components into new materials. In addition, the latest knowledge in materials engineering will be used to ensure that the product meets the static and dynamic performance of state-of-the-art materials.

FiberEUse (new value chains in a closed-loop economy based on the reuse of end-of-life fiber-reinforced composites) (FiberEUse 2020)

The project is being carried out by more than a dozen different business entities, academic institutions, design and consulting companies, from various European countries.

FiberEUse aims to integrate various innovation activities through a holistic approach to increase the profitability of composites recycling and reuse in value-added products. With new cloud-based ICT solutions applied to value chain integration, identification of new markets, analysis of regulatory barriers, life cycle assessment, the project will support the industry in its transition to a closed-loop composites economy model.

HiPerDiF (*High-Performance Discontinuous Fibre*) (HiPerDiF 2020)

The project is being carried out by a group of 11 scientists, from a variety of disciplines, associated with the University of Bristol. The team's intention is to fundamentally change the composites industry by developing and applying short-fiber technology to produce high-performance, fully recyclable composite materials.

IEA Wind Task 45 (identifying and reducing barriers to wind turbine blade recycling) (IEA Wind Task 45 2020)

The project is being implemented as part of the IEA Wind TCP, which is a forum for international cooperation among 24 countries and sponsor members sharing information and research activities to accelerate wind energy deployment. The attention of key stakeholders is focused on identifying barriers and mitigation strategies for large-scale implementation of wind turbine blade recycling solutions. The aforementioned initiative focuses on producing tangible results, such as recommended practices and guidelines that can be used by practitioners. Key topics addressed by the project include:

- technical aspects of recycling;
- shovel life cycle and value chain;
- standards and regulations that define wind turbine blade recycling activities.

ReRoBalsa (Recycling rotor blade material to recover balsa wood and plastic foam, for the production of insulation materials) (ReRoBalsa 2017)

This research project is being conducted at the Fraunhofer Institute for Wood Research.

The goal is to develop an innovative recycling technology aimed at recovering balsa wood and plastic foam, from rotor blades. An additional development of this research initiative is an innovative recycling technology for these recyclates, aimed at their use in the production of new and improved insulation and construction materials.

SusWind (faster and wider use of sustainable composite materials and technologies in wind turbine blade manufacturing) (SusWind 2019)

A project in partnership with Offshore Renewable Energy Catapult and supported by The Crown Estate and RenewableUK, SusWIND aims to develop and demonstrate viable ways to recycle composite wind turbine blades, as well as to explore the use of sustainable materials and processes in the development of composites necessary for wind blade manufacturing and in their innovative design to meet the requirements of the future.

In addition, there are still a number of projects that focus on end-of-life composites in general, such as:

Ecobulk (a project that demonstrates the applications of recycled composites in the furniture, automotive and construction industries) (Ecobulk 2020)

Ecobulk is made up of different organizations with different visions and expertise, which have joined forces to create a new cyclical value chain for wind blade composites as well. The initial step is to process GFRP (Glass Fiber-Reinforced Polymers) composites, and then plug the resulting components into material cycles, supplemented with small amounts of virgin materials. The project mainly deals with the specific case of a material based on wind turbine blade waste, but the same process works well with composite waste from other industries. The patented process developed by Conenor is flexible in terms of both the end product and the source material. Successfully produced extruded profiles to date include both hollow and filled, in a whole range of sizes and shapes. Owing to recent work done with Ecobulk partner Aimplas, it has been possible to produce, from some of these materials, pellets that can be used for injection molding. Initial cost projections for the materials are quite encouraging. Prices are in a similar range, and in some cases are even competitive with similar products on the market. These projections do not take into account economies of scale, which can also significantly reduce production costs. Cost comparisons such as for wood decking do not take into account the higher maintenance costs of wood and its lower durability compared to Ecobulk materials, which are naturally weather resistant and can be color injected during production.

Recy-composite (Recycling composite materials: a transboundary approach in a circular economy) (Recy-composite 2020)

The partners in this transboundary project are located in France as well as Wallonia and Flanders. They have complementary skills aligned with their industrial base. In addition to their recognized knowledge of plastics processes, these partners have useful expertise in waste processing and sorting, thermochemical recycling processes, fire resistance and also forming and shaping of composite materials. Each center, with its own specificities, adds real value to the joint development of transboundary activities. The project deals with the management of composites through: material recovery, thermochemical recycling (pyrolysis, solvolysis) and only as a last resort energy recovery. It is a practical example of the implementation of a closed-loop economy, with the goal of efficient use of resources and reducing the environmental impact of goods/products throughout their life cycle.

Applied research is conducted on both thermoset composite materials used in production and end-of-life composite materials. In terms of industrial transfer, the economic aspect will be taken into account to select technological solutions for recycling, in accordance with the waste treatment hierarchy described in the regulations. The project addresses specialty chemistry to propose value-added recycled products to the market.

Composite recycling generally consists of two stages (Bennet et al. 2021): reclamation and reprocessing.

The reclamation stage involves the recovery of material from the original composite, which can be used in secondary applications. In most composite recycling processes, this stage mainly focuses on recovering high-value material from the composite (fibers), but some processes are also capable of recovering resin components or energy that can power subsequent processes. Once the fibers are reclaimed, they need additional processing into a form that allows them to be used in another product.

There are four main processes for regeneration of composites (with different levels of technological readiness) (Bennet et al. 2021):

- mechanical – grinding and milling: large scale; low cost; recyclate is a mixture of components; use as a filler (when derived from thermoset composites), as a re-extrusion material (when derived from thermoplastic composites) and in the cement industry as an energy source and clinker component (only for fiberglass composites) (Oliveux et al. 2015),
- thermal:
  - conventional pyrolysis (SusWind 2019) – energy intensive; recovery of oils from resins; limited scale; relatively high cost; suitable for recovery of carbon fiber (it retains up to 90% of mechanical properties); currently not suitable for glass fiber (low product quality),
  - fluidized bed pyrolysis (Pickering 2006; Pimenta & Pinho 2012) – expensive; energy intensive; limited scale; produces good quality and pure carbon fiber (up to 70%–80% of retained mechanical properties),
  - microwave pyrolysis (SusWind; Boulanghien et al. 2015)– expensive; available on a very small scale; produces good quality carbon fibers (up to 75% retention of mechanical properties); lower energy intensity than other thermal methods,
  - thermolysis (Bennet et al. 2021) (superheated steam pyrolysis) – energy intensive; produces high-quality carbon fibers (more than 90% retention of mechanical properties); very small scale,
- chemical: solvolysis (SusWind 2019; Kao et al. 2011) (various solvents, temperature and pressure, catalysts) – high cost; small-scale availability; energy intensive; yields high-quality glass fibers (retain up to 70% of mechanical

properties) and carbon fibers (retain up to 90% of mechanical properties). The process includes:

- high-temperature and high-pressure solvolysis (SusWind 2019; Keith et al. 2016) – high cost; limited availability; high energy intensity; corrosive; produces good quality and pure carbon fibers,
- low temperature (SusWind 2019) (up to 200°C) and low pressure (atmospheric pressure), catalytic solvolysis – less energy intensive than the variant above; requires the use of acids, the sustainable disposal of which is problematic; produces good quality and pure carbon fibers and epoxy monomers.

With RecyclableBlade technology, Siemens Gamesa (2021) is able to separate and recycle wind blade materials for use in new applications. The RecyclableBlade is manufactured in the same way as the standard blade and is based on the same IntegralBlade® manufacturing process. The only difference is the use of a new type of resin that allows it to be effectively separated from other components at the end of the blade life. The process can be described most briefly in the following steps:

1. Disassembly of end-of-life blades from the turbine and their preparation for the recycling process.
2. Leaching with a diluted, heated acid solution of the blade immersed in it – separating the resin from the fiberglass, plastic, wood and metals.
3. Recovery of separated components from the solution and preparation for secondary use, i.e., rinsing, drying.
4. Reuse of recovered components in new products (guided by the technical properties of the components), i.e., in the automotive industry or in consumer goods, such as carrying cases and cases for flat screens.

- Electrochemical (SusWind; Mativenga et al. 2016) – high cost; high energy consumption; small-scale availability; yields fiberglass of useful quality.

The reprocessing stages involve converting recovered materials into secondary material for use in other applications. Recyclate in ground form is used as filler. Other forms of recyclate (shredded fibers, nonwoven mats, pellets) require post-processing to ensure that it can be reused. The aforementioned materials can be used in typical composite manufacturing processes, such as resin infusion, injection molding, compression molding and thermoplastic blending (IEA Wind Task 45 2020).

This group of processes also includes the reuse of rotor blades made of thermoset composites (reusing thermosets). The products of such mechanical reprocessing are already available on the market. The development of this approach to the use of fiber-reinforced plastics is due to its moderate cost.

The reuse of wind turbine rotor blade sections can be observed in civil engineering and construction. This approach has the least environmental impact of all the recycling methods described above. The best evidence of this can be seen in the implementation of projects such as Re-Wind and Superuse Studios NL BV:

Re-Wind is a project led by an interdisciplinary research team consisting of experts from the City University of New York, Georgia Institute of Technology, University College Cork and Queen's University Belfast. This project addresses the concepts of reusing rotor blades for various purposes. It emphasizes the need to know their mechanical properties, especially their strength (coming from different locations, from different manufacturers), especially if they are to be used as structural elements (such as a bridge, roof or electric pole). It stresses that most of the solutions that have been developed to date do not offer the scalability required for mass demand in the future, and postulates that there is a need to create solutions that can be easily modified or scaled for larger blades.

In contemporary literature, many researchers indicate repurposing as an environmentally, socially and economically sustainable option for decommissioning composite wind turbine blades. Repurposing is preferable to material recovery, waste-to-energy or landfilling within a closed-loop economy paradigm, and may offer additional social benefits, compared to the other options mentioned above. Ongoing research around the world is aimed at establishing applications for the reuse of wind turbine blade structures in civil engineering infrastructure, under the assumption that advanced composite materials can be an attractive alternative to conventional infrastructure materials (e.g., steel, reinforced concrete). Such views are increasingly frequent (Alshannaq et al. 2019; Nie et al. 2019; Bank et al. 2019).

The Dutch company Superuse Studios (Superuse 2020) uses wind blades in such projects as: playgrounds, bicycle sheds, benches and other outdoor facilities.

In Poland, it is worth mentioning the work carried out by the company Anment (2021), which deals with projects for the implementation of pedestrian and bicycle footbridges (such as in Szprotawa (Figure 4.3)), and also intends to create parking shelters, bus stops, benches or swings (Figure 4.4), using decommissioned wind blades. This company claims to have *"developed the technology for recovering*

FIGURE 4.3 Footbridge prototype (Anment 2021).

FIGURE 4.4   Bus stop design (Anment 2021).

carbon fibers from wind turbine blades" and that they "*are the only company in the world that can recover fibers up to 8 meters long.*" In addition, this company is participating, together with the Warsaw University of Technology, in research on the use of recycled fibers for the production of laminates. The university conducts strength tests on recycled carbon fibers and laminates created from them, commissioned by Anment. In collaboration with Rzeszow University of Technology, the company also intends to build bridges and footbridges, using wind turbine blades.

From recycled propellers, GP Renewables Group (2021) can manufacture all kinds of unique furniture suitable for urban, hotel or private spaces, while resistant to changing weather conditions. These are massive, although looking very light and modern, structures that are difficult to destroy or rearrange. The designers of this company also have solutions for pedestrian and bicycle bridges, and there are plans for bridges for automobile traffic, using wind blades (Figure 4.5).

FIGURE 4.5   Outdoor furniture made from wind turbine propellers (Wings for Living 2021).

Thornmann Recycling Ltd. was the first company in the country to start recycling fiberglass and carbon fiber, offering composite recyclates for many reuses (Thornmann Recycling 2021).

Some examples of already implemented concepts using wind blades are shown in Figures 4.6–4.8:

To sum up, it can be said that the wind energy sector in the broadest sense, in cooperation with scientific and research institutions and other external partners, has undertaken the task of managing even such resistant-to-processing materials as composites, which build decommissioned rotor blades of wind turbines. The experience gained so far in this area indicates the correct trend in the development of methods for managing such blades, consisting in moving away from their storage or incineration and accepting, increasingly, physical and chemical processes aimed

FIGURE 4.6   A bicycle shelter in Aalborg, Denmark (WindEurope 2020).

FIGURE 4.7   Wind blades as electric poles (Re-Wind Network 2022).

FIGURE 4.8   Playground (Superuse 2020).

at recovering components for reuse. The reuse of wind turbine rotor blade sections in civil engineering and construction is already entering the economy and finding support in the cultural and social space. This approach has the least environmental impact of all the recycling methods described above.

These processes, combined with the pretreatment of substrates, are in line with the waste management hierarchy, meet the principles of a closed-loop economy and bode well for the future.

5 Summary and Conclusions

At present, Poland, Europe and the world are faced with the enormous challenge of the inevitable energy transition. On the one hand, there is the task of ensuring energy security and rampant energy prices; on the other hand, the state of environmental degradation and climate change require urgent changes and decisions. For the ongoing discussion of the energy transition, including the share of wind power in the Polish energy mix and its impact on human health and the state of the environment, the following conclusions are particularly important:

- To date, the development of energy, based on renewable sources, and the prospects for its further development indicate that the appropriate pace and form of energy transition are not possible without a significant share of RES, especially onshore wind energy.
- The key law for the Polish onshore wind power industry regarding "investments on wind power plants" (Distance Act 2016) functions together with other legislation and the country's energy and environmental policy. The key element of the law, without which there will be no further development of onshore wind energy, is the reduction of the required minimum distance of a wind power plant from residential buildings.
- In many countries, the minimum distances between wind turbines and human residences that are taken into account are only in the nature of recommendations and have not been introduced by common law. In many cases, local governments and communities have the authority to decide on the location of wind farms. In addition, it is pointed out that the applicable guidelines are not rigid in nature and should be applied flexibly, examining each case individually.
- According to the World Health Organization (WHO), wind energy is associated with fewer negative health impacts than other forms of traditional power generation, and will even have positive health effects by reducing emissions.
- The results of expert field measurements of noise around wind farms indicate that the minimum distance in Poland, acting as a warning light, may be 500 m. However, the final distance should always be determined by calculation methods and verified by field measurements.
- In many publications, the need for public consultation from the earliest stages of investment, including extensive information campaigns on the impact of wind power on human health, is expressed directly and indirectly – this reduces the concerns of local community members.

DOI: 10.1201/9781003456711-6

- Wind turbines should be treated like any other industrial noise source. The extent of their acoustic impact does not differ from the impact of common anthropogenic noise sources. In assessing this noise, the important issue is not its perception but rather its level and nature.
- With regard to the sounds emitted by wind turbines, the vast majority of scientists agree that there is no conclusive evidence that noise, including infrasound, emitted by wind turbines, has a negative impact on human health or well-being.
- The sound produced during turbine operation is perceived differently by people, depending on their sensitivity and attitudes to the presence of this installation. The dominant factor creating public attitudes toward wind power is the audible part of the acoustic spectrum, coming from turbine operation.
- The overwhelming majority of studies indicate a significant decrease in the annoyance of acoustic impact with increasing distance from the turbine. The results of these studies also indicate that there is no effect of audible noise on the clinical changes of the subjects, as well as other potential variables that may have an effect but are not directly related to this noise.
- Observed levels of the infrasound noise from wind turbines are lower or comparable to the noise associated with typical natural sources of infrasound (e.g., wind, waves, lightning, heavy rain) commonly found in nature, and the infrasound noise accompanying humans in daily living activities (e.g., vehicles, loudspeakers, motors, household appliances, airplanes).
- To cause annoyance, the frequency of shadow flicker should be above 2.5 Hz, which, for a three-bladed turbine, would entail a figure of 50 revolutions per minute. Contemporary wind turbines rotate at between 3 and 20 revolutions per minute, giving flicker frequencies in the shadow area of 0.15–1 Hz. Significant health effects that can occur due to flicker are found only from a frequency of 3 Hz, which are not emitted by contemporary turbines.
- To minimize the effects of light reflections, it is recommended to use wind turbines painted with special paints and made of materials that reduce the reflection of light rays.
- There is also no risk to human health from wind turbines, for electromagnetic and vibrational impacts, with cardinal precautions. The current methodology for determining the degree of maximum intensity of negative impacts of wind turbines on human health, as well as acceptable standards for electromagnetic fields and vibrations, provide an adequate level of safety.
- Increasingly frequent in the wind power industry are research, implementation and commercial ventures aimed at using decommissioned out-of-service rotor blades and the components that build them in accordance with the principles of closed-cycle economy.
- There is a need for modifications to the legislation and guidelines governing the wind power industry and its various impacts on the environment, including human health and well-being.

References

Accelerating Wind Turbine Blade Circularity, May 2020. https://windeurope.org/wp-content/uploads/files/about-wind/reports/WindEurope-Accelerating-wind-turbine-blade-circularity.pdf

Ackermann T., *Wind power in power systems*, John Wiley and Sons, Chichester, 2009.

Acoustica AS, *The noise measurements*, Report P4.010.97, Vestas report 943111.R4, Denmark, 1997.

Act of June 14, 1960, Code of Administrative Procedure (Journal of Laws of 2021, item 735, as amended), CAP - in Polish.

Act of March 21, 1985 on public roads (Journal of Laws 2021, item 1376, as amended) - in Polish.

Act of January 12, 1991 on local taxes and fees (Journal of Laws of 2019, item 1170, as amended), Tax Act - in Polish.

Act of September 28, 1991 on forests (Journal of Laws 2022, item 672) - in Polish.

Act of July 7, 1994, Construction Law (Journal of Laws of 2021, item 2351, as amended), Construction Law - in Polish.

Act of February 3, 1995 on the protection of agricultural and forest land (Journal of Laws 2021, item 1326) - in Polish.

Act of April 10, 1997 - Energy Law (Journal of Laws of 2021, item 716, as amended), Energy Law - in Polish.

Act of April 27, 2001. - Environmental Protection Law (Journal of Laws 2021, item 1973, as amended), Environmental Protection Law - in Polish.

Act of July 3, 2002. - Aviation Law (Journal of Laws of 2022, item 1235) - in Polish.

Act of March 27, 2003 on spatial planning and development (Journal of Laws 2022, item 503), SPD Act - in Polish.

Act of March 11, 2004 on tax on goods and services (Journal of Laws 2022, item 931), VAT Act - in Polish.

Act of April 16, 2004 on Nature Protection (Journal of Laws 2022, item 916), Law on Nature Protection - in Polish.

Act of July 28, 2005 on spa treatment, spas and areas of spa protection and spa municipalities (Journal of Laws 2021, item 1301) - in Polish.

Act of October 3, 2008 on providing information on the environment and its protection, public participation in environmental protection and environmental impact assessments (Journal of Laws of February 13, 2020, item 283) - in Polish.

Act of December 6, 2008 on excise tax (Journal of Laws 2022, item 143), Excise Act - in Polish.

Act of February 20, 2015 on Renewable Energy Sources (Journal of Laws of 2021, item 610, as amended), RES Act - in Polish.

Act of April 24, 2015 on amending certain laws in connection with strengthening landscape protection tools (Journal of Laws of 2015, item 774), Landscape Law - in Polish.

Act of May 20, 2016 on investments in wind power plants, Journal of Laws. 2016, item 961 - In Polish.

Act of July 20, 2017. - Water Law (Journal of Laws 2021, item 2233, as amended) - in Polish.

Act of June 7, 2018 amending the Law on Renewable Energy Sources and certain other laws (Journal of Laws of 2018, item 1276 - in Polish.

Act of July 15, 2020 on amending the Act on the principles of development policy and certain other acts (Journal of Laws of 2020, item 1378) - in Polish.

Act of November 17, 2021 on the compensation of revenues lost by municipalities in 2018 due to the change in the scope of taxation of wind power plants (Journal of Laws of 2022, item 30) - in Polish.

Act on the Protection and Care of Monuments of July 23, 2003. (Journal of Laws of 2022, item 840) - in Polish.

Aircraft Impact, *International Council for research and innovation in building and construction*, report COIB 14421/2017, 2017.

Alshannaq A., Scott D., Bank L., Bermek M., Gentry R., *Structural re-use of de-commissioned wind turbine blades in civil engineering applications*, In Proceedings of the 34th ASC Technical Conference, Atlanta, GA, 2019.

Alves-Pereira M., Branco N.C., Industrial Wind Turbines, Infrasound and Vibro-Acoustic Disease (VAD), Press Release 31, 2007.

Anment._https://spidersweb.pl/2021/12/zuzyte-turbiny-wiatrowe-recykling-problem.html, 2021.

Aris A., Yiannis K., Charilaos T., et al. Extremely low frequency electromagnetic field exposure measurement in the vicinity of wind turbines, *Radiat Prot Dosimetry* 189(3), 395–400, 2020.

Australian Wind Energy Association, *The Electromagnetic Compatibility and Electromagnetic Field Implications for Wind Farming in Australia*, Report to Australian Government, Australian Greenhouse Office, 2004.

Badora K., Location of wind farms in the southern part of Opolskie Province versus natural and landscape conditions [in Polish], *Inżynieria Ekologiczna* 23, 97–107, 2010.

Badora K., *Społeczna percepcja energetyki wiatrowej na przykładzie farmy wiatrowej Kuniów* [Public perception of wind energy on the example of Kuniów wind farm - In Polish] 2017, Proceedings of ECOpole 11, 2017.

Badora K., *Zalecenia w zakresie uwzględnienia wpływu farm wiatrowych na krajobraz w procedurach ocen oddziaływania na środowisko* [Recommendations for considering the influence of wind farms on the landscape in environmental impact assessment procedures - In Polish], Generalna Dyrekcja Ochrony Środowiska, 2017.

Bakoń T., *Zapobieganie i usuwanie oblodzenia w elektrowniach wiatrowych* [Prevention and removal of icing in wind turbines - in Polish], elektro.info 9, 2013. https://www.elektro.info.pl

Bank L.C., Arias F.R., Gentry T.R., Al-Haddad T., Tasistro-Hart B., Chen J.F., Structural analysis of FRP parts from waste wind turbine blades for building reuse applications, In *Advances in engineering materials, structures and systems: innovations, mechanics and applications*, pp. 1520–1524, CRC Press/Balkema, London, 2019.

Baranowski A., Borowski S., Lubocka-Hoffmann M., Marć-Pieńkowska J., Mikołajczak J., Pojmański G., Rosiak I., *Farmy wiatrowe zagrożenia dla człowieka i środowiska na przykładzie Elbląga i Żuław Wiślanych* [Wind farms as threats to humans and the environment on the example of Elblag and Żuławy Wiślane - in Polish], Wydawnictwo Uczelniane Uniwersytetu Techniczno-Przyrodniczego, Bydgoszcz, 2014.

Barclay C., Wind Farms - *Distance from housing*, Library of House of Commons SN/SC/5221, 2012.

Baring-Gould I., Cattin R., Durstewitz M., Hulkkonen M., Krenn A., Laakso T., Lacroix A., Peltola E., Ronsten G., Tallhaug L., Wallenius T., IEA wind recommended practice 13: wind energy in cold climates, IEA Task-19, 2012.

Basner M., Muller U., Elmenhorstm E., Single and combined effects of air, road, and rail traffic noise on sleep and recuperation, *Sleep* 34, 11–23, 2011.

Bennet L., Hailey J., Lomoro P., Fitzgerald A., Fuller J., Lightfoot J., Velenturf A., Trifonova K., *Sustainable Decommissioning: Wind Turbine Blade Recycling*, report from Phase 1 of the Energy Transition Alliance Blade Recycling Project, March 2021.

Betakova V., Vojar J., Sklenicka P., Wind turbines location: how many and how far?, *Applied Energy* 151, 23–31, 2015.

Bilski B., Factors influencing social perception of investments in the wind power industry with an analysis of influence of the most significant environmental factor - exposure to noise, *Polish Journal of Environmental Studies* 21(2), 289–295, 2012.

Biswas S., *Modelling of ice throw and noise from wind turbines, earth & space science*, In A Dissertation Submitted to the Faculty of Graduate Studies in Partial Fulfillment of the Requirements, 2021.

Bolwig S., Bolkesjø T.F., Klitkou A., Lund P.D., Bergaentzlé C., Borch K., Olsen O.J., Kirkerud J.G., Chen Y.K., Gunkel P.A., Skytte K., Climate-friendly but socially rejected energy-transition pathways: the integration of techno-economic and socio-technical approaches in the Nordic-Baltic region, *Energy Research & Social Science* 67, 2020.

Bonsma I., Gara N., Howe B., McCabe N., *An investigation into short-term fluctuations in amplitude modulation of wind turbine noise: Preliminary results*, In Proceedings of the 7th International Conference on Wind Turbine Noise, Rotterdam, The Netherlands, 2–5 May 2017.

Boulanghien M., Da Silva S., Berthet F., Bernhart G., Soudais Y., Using steam thermolysis to recycle carbon fibres from composite waste, *JEC Composites Magazine* 100, 68–70, 2015.

Bresden R.E., Drapalik M., But B., Understanding and acknowledging the ice throw hazard - consequences for regulatory frameworks, risk perception and risk communication, *Journal of Physics: Conference Series* 926, 012001, 2017.

British Epilepsy Society, *Wind turbines and photosensitive epilepsy.* https://epilepsysociety. org.uk, 2020.

Bullmore A., Adcock J., Jiggins M., Cand M., *Wind farm noise predictions and comparison with measurements*, In Third International Meeting on Wind Turbine Noise, Aalborg, Denmark, 2009.

Carlsson V., Measuring routines of ice accretion for wind turbine applications. The correlation between production losses and detection of ice, *WINTERWIND* 2011, https://windren. se/WW2011/32_Skekraft_Carlsson_icing_production.pdf

Cattin R., Kunz S., Heimo A., Russi G., Russi M., Tiefgraber M., Wind turbine ice throw studies in the Swiss Alps, *European Wind Energy Conference and Exhibition* 2007, 3, 1588–1592, 2014.

CETEC. https://www.vestas.com/en/media/company-news/2021/new-coalition-of-industry-and-academia-to-commercialise-c3347473#!NewsView, 2021.

Chapman S., St. George A., How the factoid of wind turbines causing "vibroacoustic disease" came to be "irrefutably demonstrated", *Australia New Zealand Journal of Public Health* 33, 244–249, 2013.

Chojnacka I., Nowicki M., *Omówienie do wyroku TK z dnia 22 lipca 2020 r.* [Discussion to the ruling of the Constitutional Tribunal of July 22, 2020 - in Polish], *K 4/19, ZNSA*, nr 4, 123–128, 2020.

Chylarecki P., Kajzer K., Polakowski M., Wysocki D., Tryjanowski P., Wuczyński P.A., *Wytyczne dotyczące ocen oddziaływania elektrowni wiatrowych na ptaki.* [Guidelines for assessing the impact of wind turbines on birds - in Polish] *PROJEKT*, Generalna Dyrekcja Ochrony Środowiska, 2011.

Circular Wind Hub. https://www.echt.community/the-circular-wind-hub/, 2019.

Cohen J.J., Reichl J., Schmidthaler M., Re-focussing research efforts on the public acceptance of energy infrastructure: a critical review, *Energy* 76, 4–9, 2014.

Colloca L., Tell me the truth and I will not be harmed: informed consents and nocebo effects, *The American Journal of Bioethics* 17(6), 46–48, 2017.

Council of Canadian Academies, *Understanding the Evidence: Wind Turbine Noise. Ottawa (ON): The Expert Panel on Wind Turbine Noise and Human Health*, Council of Canadian Academies, 2015.

Czyżak P., Sikorski M., Wrona A., Wiatr w żagle. *Zasada 10H a potencjał lądowej energetyki wiatrowej w Polsce* [Getting a second wind. The 10H rule and the potential of onshore wind power in Poland - in Polish], Instrat Policy Note 01, 2021.

Davy J.L., Burgermeister K., Hillman D., Wind turbine sound limits: current status and recommendations based on mitigating noise annoyance, *Applied Acoustics* 140, 288–295, 2018.

Davy J.L., Burgemeister K., Hillman D., Carlile S., A review of the potential impacts of wind turbine noise in the Australian context, *Acoustics Australia* 48, 181–197, 2020.

DecomBlades. https://decomblades.dk/

DecomTools. https://northsearegion.eu/decomtools/, 2020.

DELTA, *Nordic Environmental Noise Prediction Methods*, Nord 2000, Summary report, Lyngby, Denmark, 2002.

Dreamwind. https://www.dreamwind.dk/en/, 2020.

Directive 2002/49/EC of the European Parliament and of the Council of 25 June 2002 relating to the assessment and management of environmental noise.

Commission Directive (EU) 2015/996 of 19 May 2015 establishing common noise assessment methods according to Directive 2002/49/EC of the European Parliament and of the Council.

Directive 2014/30/EU of the European Parliament and of the Council of 26 February 2014 on the harmonisation of the laws of the Member States relating to electromagnetic compatibility.

Directive 2009/147/EC of the European Parliament and of the Council of 30 November 2009 on the conservation of wild birds.

Directive 2009/28/EC of the European Parliament and of the Council of 23 April 2009 on the promotion of the use of energy from renewable sources and amending and subsequently repealing Directives 2001/77/EC and 2003/30/EC.

Council Directive 92/43/EEC of 21 May 1992 on the conservation of natural habitats and of wild fauna and flora.

Ecobulk. https://www.ecobulk.eu/, 2020.

Ecotière D., Gauvreau B., Cotté B., Roger M., Schmich-Yamane I., Nessi M.C., *PIBE: A new French project for predicting the impact of wind turbine noise*, In Proceedings of the 8th International Conference on Wind Turbine Noise, Lisbon, Portugal, 2019.

Elektrownie wiatrowe - fakty i mity [Wind power plants - facts and myths - in Polish]. https://archiwum.myslowice.net/assets/files/entries/Elektrowniewiatrowefaktyimity.pdf   (last accessed May 2021).

Ellis G., Ferraro G., *The Social Acceptance of Wind Energy. Where We Stand and the Path Ahead*, JRC Science for Policy Report, European Commission, Brussels, 2016.

Environment Protection and Heritage Council (EPHC), *National Wind Farm Development Guidelines - Draft*, 2010.

ETIP Wind, *How Wind Is Going Circular*, 2019. ETIP Wind-How-wind-is-going-circular-blade-recycling.pdf

European Commission Decision No. SA.64713, C 511/1, 2021.

European Commission, *Guidance document: Wind energy development and Natura 2000*, EU Guidance on wind energy development in accordance with the EU nature legislation, 2011.

Evans T., Cooper J., Comparison of predicted and measured wind farm noise levels and implications for assessments of new wind farms, *Acoustics Australia* 40(1), 28–36, 2012a.

Evans T., Cooper J., *Influence of wind direction on noise emission and propagation from wind turbines*, In Proceedings of Acoustics, 2012b.

FiberEUse. https://www.fibereuse.eu/, 2020.

Flaga Ł., Flaga A., Szeląg A., *O planowaniu przestrzennym i uwarunkowaniach środowiskowych siłowni i farm wiatrowych dużej mocy* [On spatial planning and environmental conditions of high-power wind power plants and wind farms - in Polish], *Inżyn. i Bud.* 12, 40, 2018.

Fortin P., Rideout K., Copes R., Bos C., *Wind turbines and health*, National Collaborating Centre for Environmental Health at the British Columbia Centre for Disease Control, 2013.

Fredianelli L., Carpita S., Licitra G., A procedure for deriving wind turbine noise limits by taking into account annoyance, *Science of the Total Environment* 648, 728–736, 2019.

Freiberg A., Schefter C., Hegewald J., Seidler A., The influence of wind turbine visibility on the health of local residents: a systematic review, *International Archives of Occupational and Environmental Health* 92(5), 609–628, 2019.

Gaj P., Błaszczak J.R., *Pomiary drgań przy użyciu niestacjonarnego systemu monitoringu turbin wiatrowych* [Vibration measurements using a non-stationary wind turbine monitoring system - in Polish], *Cieplne Maszyny Przepływowe* 143, 23–32, 2013.

Gamboa G., Munda G., The problem of windfarm location: a social multi-criteria evaluation framework, *Energy Policy* 35, 1564–1583, 2007.

GEO Renewables S.A., *Korytnica Wind Farm Non-Technical Summary*, Wind farm Korytnica, Poland, 2014.

Giuffrida L.G., Penna M., De Luca E., Nardi C., Colosimo A., Krug M., *Screening of technical and non-technical regulations, guidelines and recommendations*, Policy Lessons and Guiding Principles and Criteria for Fair and Acceptable Wind Energy, WinWind, 2019.

Golec M., Golec Z., Cempel C., *Hałas Turbiny Wiatrowej VESTAS V80 podczas eksploatacji* [Noise of the VESTAS V80 Wind Turbine during operation - in Polish], *Diagnostyka* 1 (37), 2006.

Gortsas T., Triantafyllidis T., Kudella P., Zieger T., Ritter J., *Low-frequency micro-seismic radiation by wind turbines and it's interaction with acoustic noise emission*, In Proceedings of the 7th International Conference on Wind Turbine Noise, Rotterdam, Netherlands, 2017.

Gouvernement Walon, Arrêté du Gouvernement wallon portant conditions sectorielles relatives aux parcs d'éoliennes d'une puissance totale supérieure ou égale à 0,5 MW, 2014.

GP Renewables Group. https://gp-renewables.energy/blades-recycling-2/

Gryz K., Karpowicz J., Ekspozycja na pole elektromagnetyczne w elektrowniach wiatrowych [Exposure to electromagnetic fields in wind power plants - in Polish], *Bezpieczeństwo Pracy* 7, 10–13, 2016.

Gschiel E., Umweltmedizinische Begutachtung am Beispiel Windkraft Amt der Burgenland Landersregierung, 2014.

Guarnaccia C., Mastorakis N.E., Quartieri J., Wind turbine noise: theoretical and experimental study, *International Journal of Mechanics* 5(3), 129–137, 2011.

Haac R., Darlow R., Kaliski K., Rand J., Hoen B., In the shadow of wind energy: predicting community exposure and annoyance to wind turbine shadow flicker in the United States, *Energy Research & Social Science* 87, 102471, 2022.

Hajto M., Cichocki Z., Bidłasik M., Borzyszkowski J., Kusmierz A., Constraints on development of wind energy in Poland due to environmental objectives. Is there space in Poland for wind farm siting?, *Environmental Management* 59, 204–217, 2017.

Hansen C.H., Doolan C.J., Hansen K.I., Wind farm noise: *measurement, assessment and control*, First Edition, John Wiley & Sons Ltd., United Kingdom, 2017.

Hansen K.L., Nguyen P., Zajamšek B., Catcheside P., Hansen C.H., Prevalence of wind farm amplitude modulation at long-range residential locations, *Journal of Sound and Vibration* 455, 136–149, 2019.

Haugen K.B.M., International Review of Policies and Recommendations for Wind Turbine Setbacks from Residences, Minnesota Department of Commerce: Energy Facility, 2011.

Helbin J., *Środowiskowe czynniki fizyczne wpływające na organizm człowieka* [Environmental physical factors affecting the human body - in Polish], Wybrane problemy higieny i ekologii człowieka, Wydawnictwo UJ. Kraków, 2008.

HiPerDiF._https://www.researchgate.net/project/HiPerDiF-High-Performance-Discontinuous-Fibre, 2020.

Hojan E., *Protetyka słuchu* [Hearing care - in Polish], Wydawnictwo Naukowe UAM, 2014.

https://www.windturbinenoise.eu/ (last accessed May 2021).

Hua L., Zhu X., Hua C., Chen J., Du Z., Wind turbines ice distribution and load response under icing conditions, *Renew Energy* 113, 608–19, 2017.

Huebner G., Pohl J., Hoen B., Firestone J., Rand J., Elliott D., Haac R., Monitoring annoyance and stress effects of wind turbines on nearby residents: a comparison of U.S. and European samples, *Environment International* 132(105090), 1–9, 2019.

Ibrahim G.M., Pope K., Muzychka Y.S., Effects of blade design on ice accretion for horizontal axis wind turbines, *Journal of Wind Engineering & Industrial Aerodynamics* 173, 39–52, 2018.

IEA, *Net Zero by 2050: A Roadmap for the Global Energy Sector*, International Energy Agency, Paris, 2021.

IEA Wind Task 45. https://iea-wind.org/task45/, 2020.

IEC 61400-11 Ed.3.0. *Wind Turbines-Part 11: Acoustic Noise Measurement Techniques*, International Electrotechnical Commission, Geneva, Switzerland, 2012.

India Centre for Science and Environment, *EIA Guidelines Wind Power Sector*, 2013.

Ingielewicz R., Zagubień A., *Pomiar hałasu infradźwiękowego wokół farmy wiatrowej* [Measurement of infrasound noise around a wind farm - in Polish], *Wydawnictwo PAK* 59(7), 725–727, 2013.

Ingielewicz R., Zagubień A., Infrasound noise of natural sources in environment and infrasound noise of wind turbines, *Polish Journal of Environmental Studies* 23, 1323–1327, 2014a.

Ingielewicz R., Zagubień A., *Tłumienie gruntu w analizach akustycznych farm wiatrowych* [Ground attenuation in acoustic analyses of wind farms - in Polish], *Wydawnictwo PAK* 60(2), 118–121, 2014b.

Distribution Network Operation and Maintenance Manual - in Polish, PGE Dystrybucja - Operator Sieci, aktualizacja na 27.08.2020 r.

Instytut OZE Sp. z o.o., *Delimitacja obszarów potencjalnej lokalizacji dużej energetyki wiatrowej na terenie województwa warmińsko-mazurskiego* [Delimitation of areas of potential location of large wind energy in the Warminsko-Mazurskie Province - in Polish], s.134, 2013.

IPCC, Climate Change 2022: Impacts, Adaptation and Vulnerability Summary for Policy Makers, Intergovernmental Panel on Climate Change, Bonn, 2022a.

IPCC, Climate Change 2022: Impacts, Adaptation and Vulnerability-Technical summary, Intergovernmental Panel on Climate Change, Bonn, 2022b.

IRENA, Future of wind: Deployment, investment, technology, grid integration and socio-economic aspects, A Global Energy Transformation Paper, 2019a.

IRENA, Global energy transformation: A roadmap to 2050 (2019 edition), International Renewable Energy Agency, Abu Dhabi, 2019b.

IRENA, Innovation outlook: Thermal energy storage, International Renewable Energy Agency, Abu Dhabi, 2020.

IRENA, World Energy Transitions Outlook: 1.5°C Pathway, International Renewable Energy Agency, Abu Dhabi, 2021. IRENA_World_Energy_Transitions_Outlook_2022.pdf

ISO-2631:1997, Mechanical vibration and shock - Evaluation of human exposure to whole-body vibration.

Jacobsen J., Danish guidelines on environmental low frequency noise, infrasound and vibration, *Journal of Low Frequency Noise, Vibration and Active Control* 20(3), 141–148, 2001.

Jagoda J., *Sądowa ochrona samodzielności jednostek samorządu terytorialnego* [Judicial protection of self-government units' independence - in Polish], Wolters Kluwer Polska, 2011.

Japanese Government Ministry of the Environment, Summary Report of Working Group Regarding Basic Concept on Environmental Impact Assessment Related to Wind Power Generation Facilities, 2010.

Joyce Lee, Feng Zhao, Global Wind Report 2021, Brussels, Belgium. Published: 25 March 2021. https://gwec.net/wp-content/uploads/2021/03/GWEC-Global-Wind-Report-2021.pdf

Kao C. C., Ghita O., Evans K. E., Oliveux G., Mechanical characterisation of glass fibres recycled from thermosetting composites using water-based solvolysis process, *International Conference on Composite Materials* 18, 1–5, 2011.

Kariniotakis G., *Renewable energy forecasting - from models to applications*. Woodhead Publishing Elsevier, United Kingdom, 2017.

Karpat E., Karpat F., Electromagnetic compatibility (EMC) in wind turbines, *International Journal of Industrial Electronics and Electrical Engineering* 4(6), 90–93, 2016.

Keikko T., Isokorpi J., Reivonen S., Ruoho T., Korpinen L., Magnetic field measurements and calculations with 20kV underground power cables, *Transaction on Modelling and Simulation* 21, 1999.

Keith M. J., Oliveux G., Leeke G. A., *Optimisation of solvolysis for recycling carbon fibre reinforced composites*, In *ECCM 2016 - Proceedings of the 17th European Conference on Composite Materials*. 17, 2016.

Kepel A., Ciechanowski M., Jaros R., *Wytyczne dotyczące oceny oddziaływania elektrowni wiatrowych na nietoperze* [Guidelines for assessing the impact of wind turbines on bats - In Polish] *PROJEKT*, Generalna Dyrekcja Ochrony Środowiska, 2013.

Kirpluk M., *Podstawy akustyki* [Fundamentals of acoustics - in Polish], NTL-M. Kirpluk, 2017.

Kistowski M., *Propozycja metodyczna oceny środowiskowych uwarunkowań lokalizacji farm wiatrowych w skali regionalnej* [Methodological proposal for assessing the environmental conditions of wind farm locations on a regional scale - in Polish], *Przegląd Geograficzny* 84(1), 5–22, 2012.

Knopper L.D., Ollson C.A., Health effects and wind turbines: a review of the literature, *Environmental Health* 10(1), 1–10, 2011.

Knopper L.D., Olsson C.A., McCallum L.C., Whitfield Aslund M.L., Berger R.G., Souweine K., McDaniel M., Wind turbines and human health, *Front Public Health* 2, 63, 2014.

Kolendow Ł., *Regionalna analiza przydatności terenów do rozwoju energetyki wiatrowej świetle wybranych uwarunkowań* [Regional analysis of the suitability of sites for wind energy development in light of selected conditions - in Polish], *Ekonomia i Środowisko* 2(56), 175–189, 2016.

Komisja Europejska, Dyrekcja Generalna ds. Środowiska, *Wytyczne dotyczące inwestycji sektora energetyki wiatrowej i przepisów UE w dziedzinie ochrony przyrody* [Guidelines for wind power sector investment and EU nature conservation regulations - in Polish], Urząd Publikacji, 2021.

Konstytucja Rzeczpospolitej Polskiej z dnia 2 kwietnia 1997 r. (Dz. U. z 1997 r. Nr 78, poz. 483, ze zm.) [Constitution of the Republic of Poland of April 2, 1997 (Journal of Laws of 1997, No. 78, item 483, as amended - in Polish], Konstytucja RP.

Koppen E., Gunuru M., Chester A., *International legislation and regulations for wind turbine shadow flicker impact*, In 7th International Conference on Wind Turbine Noise, Holandia, 2017.

Kowalczyk R., Raport o oddziaływaniu na środowisko przedsięwzięcia polegającego na budowie zespołu turbin wiatrowych "Malesowizna" wraz z infrastrukturą [Report on the environmental impact of the project involving the construction of a wind turbine complex "Malesowizna" with infrastructure - in Polish], Ecoplan. Opole, 2014.

Krohn S., Damborg S., On public attitudes towards wind power, *Renewable Energy* 16, 954–960, 1999.

Krug F., Lewke B., Electromagnetic interference on large wind turbines, *Energies* 2, 1118–1129, 2009.

Landeta-Manzano B., Arana-Landin G., Calvo P.M., Heras-Saizarbitoria I., Wind energy and local communities: a manufacturer's efforts to gain acceptance, *Energy Policy* 121, 314–324, 2018.

Lennie M., Pechlivanoglou G., Development of Ice Throw Model for Wind Turbine Simulation Software QBlade, AIAA Scitech 2019 Forum, 1–13, 2019.

Leventhall G., Low frequency noise. What we know, what we do not know, and what we would like to know, *Journal of Low Frequency Noise, Vibration and Active Control* 28(2), 79–104, 2009.

Leventhall G., Infrasound from wind turbines-fact, fiction or deception, *Canadian Acoustics* 34(2), 29–36, 2011.

Lipka J., Wytrzymałość maszyn wirnikowych [Durability of rotor machines - in Polish], WNT, Warszawa, 1967.

Lis T., Nowacki K., Bendkowska-Senator K., *Kształtowanie optymalnych warunków pracy przy występowaniu hałasu zawodowego i pozazawodowego* [Formation of optimal working conditions in the presence of occupational and non-occupational noise - In Polish], XVIII Konferencja Innowacje w zarządzaniu i inżynierii produkcji, Zakopane, 2015.

Lubośny Z., *Elektrownie wiatrowe w systemie elektroenergetycznym* [Wind power plants in the power system - in Polish], WNT, Warszawa, 2006.

Lubośny Z., *Farmy wiatrowe w systemie elektroenergetycznym* [Wind farms in the power system - in Polish], WNT, Warszawa, 2009.

Maijala P., Turunen A., Kurki I., Vainio L., Pakarinen S., Kaukinen C., Lukander K., Tiittanen P., Yli-Tuomi, T., Infrasound does not explain symptoms related to wind turbines, Finland Prime Minister's Office, Helsinki, 2020.

Makarewicz R., Is a wind turbine a point source? *The Journal of the Acoustical Society of America* 129, 579–81, 2011.

Maleńczuk W., *Przestrzenne aspekty lokalizacji energetyki wiatrowej w województwie lubelskim* [Spatial aspects of wind power localization in the Lublin Provence - In Polish], Biuro Planowania Przestrzennego, Lublin, 2009.

Marcillo O.E., Carmichael J., The detection of wind-turbine noise in seismic records, *Seismological Research Letters* 89, 1826–1837, 2018.

Massachusetts Department of Environmental Protection, *Wind Turbine Health Impact Study: Report of Independent Expert Panel*, Massachusetts, 2012.

Mativenga P. T., Shuaib N. A., Howarth J., Pestalozzi F., Woidasky J., High voltage fragmentation and mechanical recycling of glass fibre thermoset composite, *CIRP Annals - Manufacturing Technology* 65(1), 45–48, 2016.

Matuszczak K., Flizikowski J., Zagadnienia ogólne dotyczące obciążenia środowiska w cyklu życia elektrowni wiatrowej [General issues of environmental burden in the life cycle of a wind power plant - in Polish], *Postępy w Inżynierii Mechanicznej* 6(3), 35–42, 2015.

McCallum L.C., Whitfield Aslund M.L., Knopper L.D., Ferguson G.M., Ollson C.A., Measuring electromagnetic fields (EMF) around wind turbines in Canada: is there a human health concern? *Environmental Health* 13(1), 2014.

Michaud D.S., Feder K., Keith S.E., Voicescu S.A., Marro L., Than J., Guay M., Denning A., Bower T., Villeneuve P.J., Russell E., Self-reported and measured stress related responses associated with exposure to wind turbine noise, *The Journal of the Acoustical Society of America* 139(3), 1467–1479, 2016.

Ministry of the Environment, *Noise guidelines for wind farms*, Interpretation for Applying MOE NPC Publications to Wind Power Generation Facilities, Ontario, October 2008.

Moc zainstalowana (MW) [Installed capacity (MW) - in Polish] https://www.ure.gov.pl/pl/oze/potencjal-krajowy-oze/5753,Moc-zainstalowana-MW.html

Mohamed M.H., Aero-acoustics noise evaluation of H-rotor Darrieus wind turbines, *Energy* 65, 596–604, 2014.

Molnarova K., Sklenicka P., Stiborek J., Svobodova K., Salek M., Brabec E., Visual preferences for wind turbines: Location, numbers and respondent characteristics, *Applied Energy* 92, 269–278, 2012.

Moonshot Circular Wind Farms. https://www.echt.community/moonshot/, 2020.

MOSS Computer Grafik Systeme GmbH, *WindPASS shadow*. https://www.esri.com/partners/moss-computer-grafik-a2T70000000TNZLEA4/windpass-shadow-a2d70000001AVh5AAG

Mroczek B., *Akceptacja dorosłych Polaków dla energetyki wiatrowej i innych odnawialnych źródeł energii* [Acceptance of adult Poles towards wind power and other renewable energy sources - In Polish], streszczenie raportu z badań, Pomorski Uniwersytet Medyczny, Polskie Stowarzyszenie Energetyki Wiatrowej, Szczecin, 2011.

Narodowe Centrum Badan i Rozwoju, *Fakty i mity o elektrowniach wiatrowych* [Facts and myths about wind power plants - in Polish] 2021, https://www.gov.pl/web/ncbr/fakty-i-mity-o-elektrowniach-wiatrowych.

National Association of Regulatory Utility Commissioners (NARUC) Grants & Research Department, Wind Energy & Wind Park Siting and Zoning Best Practices and Guidance fos States, 2012.

Natural Forces Wind Inc., Gaetz Brook Wind Farm Shadow Flicker Assessment Report 2013.

Nawrotek M., *Efekt migotania cienia* [Shadow flicker effect - In Polish], GLOBEnergia: Odnawialne Źródła Energii (2), 2012.

Network Code, Requirements for Generation, UE 2016/631.

Nguyen D.P., Hansen K., Zajamsek B., Human perception of wind farm vibration, *Journal of Low Frequency Noise, Vibration and Active Control* 39(1), 17–27, 2020.

Nie X.F., Fu B., Teng J.G., Bank L.C.,Tian Y., *Shear behavior of reinforced concrete beams with GFRP needles as coarse aggregate partial replacement: Full-scale experiments*, Advances in Engineering Materials, Structures and Systems: Innovations, Mechanics and Applications, 1548–1553, 2019.

Niecikowski K., Kistowski M., Uwarunkowania i perspektywy rozwoju energetyki wiatrowej na przykładzie strefy pobrzeży i wód przybrzeżnych województwa pomorskiego [Conditions and prospects for the development of wind energy on the example of the coastal zone and coastal waters of Pomorskie Province - In Polish], Fundacja Rozwoju Uniwersytetu Gdańskiego, Gdańsk, 2008.

N SEP-E 004 Standard, Electric power and signal cable lines, Design and construction - In Polish.

Oerlemans S., *An Explanation for Enhanced Amplitude Modulation of Wind Turbine Noise*, Report to Renewable UK; Technical Report; National Aerospace Laboratory, NLR: Amsterdam, Netherlands, 2011.

Oliveux G., Dandy L. O., Leeke G. A., Current status of recycling of fibre reinforced polymers: review of technologies, reuse and resulting properties, *Progress in Materials Science* 72, 61–99, 2015.

Olszowiec P., *Energetyka wiatrowa - bilans 2020 r.* [Wind energy - 2020 balance sheet - in Polish], Wiadomości Elektrotechniczne 7, 2021.

Onakpoya I.J., O'Sullivan J., Thompson M.J., Heneghan C.J., The effect of wind turbine noise on sleep and quality of life: a systematic review and meta-analysis of observational studies, *Environment International* 82, 1–9, 2015.

Ove Arup and Partners, Planning for Renewable Energy. A Companion Guide to PPS22, Stationery Office, 2004.

Palmer W.K.G., Wind turbine public safety risk, direct and indirect health impacts, *Journal of Energy Conservation* 1(1), 41–78, 2018.

Pawlaczyk-Łuszczyńska M., Zaborowski K., Dudarewicz A., Zamojska-Daniszewska, M., Waszkowska M., Response to noise emitted by wind farms in people living in nearby areas, *International Journal of Environmental Research and Public Health* 15(8), 1575, 2018.

Pawlas K., Pawlas N., Boroń M., *Życie w pobliżu turbin wiatrowych, ich wpływ na zdrowie - przegląd piśmiennictwa* [Living near wind turbines, their impact on health - a review of the literature - in Polish], *Medycyna Srodowiskowa* [*Environmental Medicine*] 15(4), 150–158, 2012.

Pedersen E., Halmstad H., *Noise annoyance from wind turbines: A review - Report 5308,* Swedish Environmental Protection Agency, 2006.

Pedersen E., Larsman P., The impact of visual factors on noise annoyance among people living in the vicinity of wind turbines, *Journal of Environmental Psychology* 28, 379–389, 2009.

Pedersen E., Persson Waye K., Perception and annoyance due to wind turbine noise - a dose - response relationship, *Journal of the Acoustical Society of America* 116, 3460–3470, 2004.

Pedersen E., Persson Waye K., Wind turbine noise, annoyance and self-reported health and well-being in different living environments, *Journal of Occupational and Environmental Medicine* 64, 480–486, 2007.

Pedersen E., Persson Waye K., Wind turbines - low level noise sources interfering with restoration? *Environmental Research Letters* 3, 1–5, 2008.

Pedersen E., van den Berg F., Bakker R., Bouma J., Response to noise from modern wind farms in The Netherlands, *Journal of the Acoustical Society of America* 126, 634–644, 2009.

Performance Specification, Vestas V150-4.0/4.2 MW 50/60 Hz, Performance Specification V150-4.0/4.2 MW 50/60 Hz, 0067-7067 V10, 2019.

Peri E., Becker N., Tal A., What really undermines public acceptance of wind turbines? A choice experiment analysis in Israel, *Land Use Policy* 99, 105113, 2020.

Piasecka I., *Badanie i ocena cyklu życia zespołów elektrowni wiatrowych* [Study and evaluation of the life cycle of wind turbine assemblies - In Polish] Rozprawa Doktorska, Politechnika Poznańska, 2014.

Pickering S.J., Recycling technologies for thermoset composite materials-current status, *Composites Part A: Applied Science and Manufacturing* 37 (8), 1206–1215, 2006.

Pierpont N., Wind turbine syndrome: a report on a natural experiment, *Santa Fe*, 294, 2009.

Pimenta S., Pinho S. T., The effect of recycling on the mechanical response of carbon fibres and their composites, *Composite Structures* 94(12), 3669–3684, 2012.

PN-B-02170:2016-12 Evaluation of the harmfulness of vibrations transmitted by the ground to buildings - In Polish, PKN, Warszawa, 2016.

PN-B-02171:2017, Assessing the impact of vibrations on people in buildings - in Polish, PKN, Warszawa, 2017.

PN-ISO 7196:2002, Acoustics - Frequency characteristics of a filter for infrasound measurements - In Polish, PKN, Warszawa 2002.

PN-ISO 9613-2:2002, Acoustics. Attenuation of sound during propagation in open space. General method of calculation - in Polish. PKN Warsaw 2002.

PN-Z 01338:2010, Acoustics - Measurement and evaluation of infrasound noise at workplaces - In Polish, PKN, Warsaw 2010.

Pohl J., Rudolph D., Lyhne I., Clausen N.E., Aaen S.B., Hübner G., Kørnøvd L., Kirkegaard J.K., Annoyance of residents induced by wind turbine obstruction lights: A cross-country comparison of impact factors, *Energy Policy* 156, 112437, 2021.

Pojmański G., *Opinia dotycząca zagrożeń związanych z eksploatacją i awariami turbin wiatrowych* [Opinion on the risks associated with the operation and failure of wind turbines - In Polish], Uniwersytet Warszawski, https://www.sndb.pl/wiatraki/files/opinia-bezpieczenstwo-a-wiatraki-g-pojmanski.pdf

Polskie Stowarzyszenie Energetyki Wiatrowej, *Stan energetyki wiatrowej w Polsce w 2016 r.* [The state of wind power in Poland in 2016 - in Polish], 2017.

Polskie Stowarzyszenie Energetyki Wiatrowej, *Wytyczne w zakresie oceny oddziaływania elektrowni wiatrowych na ptaki* [Guidelines for assessing the impact of wind turbines on birds - in Polish], Szczecin, 2008.

Poulsen A.H., Raaschou-Nielsen O., Peña A., Hahmann A.N., Nordsborg R.B., Ketzel M., Brandt J., Sørensen M., Short-term nighttime wind turbine noise and cardiovascular events: a nationwide case-crossover study from Denmark, *Environment International* 114, 160–166, 2018.

Probst F., Probst W., Huber B., *Large-scale calculation of possible locations for specific wind turbines under consideration of noise limits*, In Proc. InterNoise, The 42nd International Congress and Exposition on Noise Control Engineering, 2013.

PSEW, DWF, Baker Tilly TPA, TPA Poland, Polska Energetyka Wiatrowa 4.0 [Polish Wind Power Industry 4.0 - in Polish], 2022.

Radun J., Maula H., Saarinen P., Keränen J., Alakoivu R., Hongisto V., Health effects of wind turbine noise and road traffic noise on people living near wind turbines, *Renewable and Sustainable Energy Reviews* 157, 2022.

Rakoczy B., Ustawa o udostępnianiu informacji o środowisku i jego ochronie, udziale społeczeństwa w ochronie środowiska oraz o ocenach oddziaływania na środowisko. Komentarz [Law on providing information about the environment and its protection, public participation in environmental protection and environmental impact assessments. Commentary - in Polish], Warszawa art. 82, 2010.

Raman G., Ramachandran C.R., Aldeman R.M., A review of wind turbine noise measurements and regulations, *Wind Engineering* 40(4), 319–342, 2016.

Rand J., Hoen B., Thirty years of North American wind energy acceptance research: what have we learned? *Energy Research & Social Science* 29, 135–148, 2017.

Rausand M., Risk Assessment. Theory, Methods and Application, Wiley, 2011.

Recy-composite. https://www.recycomposite-interreg.eu/index.php/en/

Renström J., Modelling of Ice Throws from Wind Turbines, Degree Project at the Department of Earth Sciences 308, 2015.

ReRoBalsa. https://www.wki.fraunhofer.de/en/researchprojects/2017/ReRoBalsa_rotor-blades-recycling-balsa-wood-plastic-foam-for-insulation-materials.html, 2017.

Rewind._https://pl.search.yahoo.com/search?fr=mcafee_uninternational&type=E211PL105 G0&p=The+Re-Wind+Network

Robinson C.M.E., Paramasivan E.S., Taylor E.A., Morrison, A.J.T., Sanderso E.D., Study and development of a methodology for the estimation of the risk and harm to persons from wind turbines, Research Report, Health and Safety Executive RR968, 2013.

Rogers A.L., Manwell J.F., Wright S., *Wind turbine acoustic noise. Renewable Energy Research Laboratory Amherst*, University of Massachusetts, 2006.

Regulation of the Minister of Climate and Environment of September 7, 2021 on the requirements for conducting measurements of emissions (Journal of Laws. 2021 item.1710) - In Polish.

Regulation of the Minister of Climate of February 17, 2020 on ways to verify compliance with the levels of electromagnetic fields in the environment (Journal of Laws of February 18, 2020, item 258) - in Polish.

Regulation of the Minister of Economy and Labor of August 5, 2005 on occupational safety and health in work involving exposure to noise or mechanical vibration (Journal of Laws of 2005. No. 157, item 1318 - In Polish.

Regulation of the Minister of the Environment of Denmark. Bekendtgørelse om støj fra vind-møller, Bekendtgørelse nr 1284 af 15/12/2011.

Regulation of the Minister of the Environment of Denmark. Bekendtgørelse om støj fra vind-møller, Bekendtgørelse nr 135 af 07/02/2019.

Regulation of the Minister of the Environment of November 12, 2007 on the scope and manner of conducting periodic studies of electromagnetic field levels in the environment (Journal of Laws of November 12, 2007, item 1645) - in Polish.

Regulation of the Minister of the Environment of June 14, 2007 on permissible levels of noise in the environment (Journal of Laws of 2014, item 112) - in Polish.

Regulation of the Minister of Environment of July 2, 2010 on notification of installations generating electromagnetic fields (Journal of Laws No. 130, item 879) - in Polish.

Regulation of the Minister of the Environment of July 2, 2010 on the types of installations the operation of which requires notification (Journal of Laws of August 12, 2019, item 1510, consolidated text) - in Polish.

Regulation of the Minister of the Environment of October 30, 2003 on permissible levels of electromagnetic fields in the environment and ways to verify compliance with these levels (Journal of Laws of 2003, No. 192, item 1883) - in Polish.

Regulation of the Minister of Health of December 17, 2019 on permissible levels of electromagnetic fields in the environment (Journal of Laws of December 19, 2019, item.2448) - In Polish.

Regulation of the Council of Ministers of September 10, 2019 on projects that may significantly affect the environment (Journal of Laws of 2019, item 1839, as amended), EIA Regulation - in Polish.

Schmidt J.H., Klokker M., Health effects related to wind turbine noise exposure: a systematic review, *PLoS One* 9(12), 2014.

Sedaghatizadeh N., Arjomandi M., Cazzolato B., Kelso R., Wind farm noises: mechanisms and evidence for their dependency on wind direction, *Renewable Energy* 109, 311–322, 2017.

Seifert H., Westerhellweg A., Kroning J., *Risk analysis of ice throw from wind turbines*, In Proceedings of the BOREAS, Pyha, Finland 9-22, 1–9, 2003.

Shadow Flicker Impact Analysis for the Ashley Wind Energy Project, CPV Ashley Renewable Energy Company, 2010.

Shadow Variations from Wind Turbines. https://xn--drmstrre-64ad.dk/wp-content/wind/miller/windpower%20web/en/tour/env/shadow/shadow2.htm

Shohag M.A.S., Hammel E.H., Olawale D.O., Okoli O.I., Damage mitigation techniques in wind turbine blades: a review, *Wind Engineering* 41(3), 185–210, 2017.

Siemens Gamesa pioneers wind circularity: launch of world's first recyclable wind turbine blade for commercial use offshore, 2021. https://www.siemensgamesa.com/-/media/siemensgamesa/downloads/en/sustainability/environment/siemens-gamesa-20210901-recycableblade-infographic-finalen.pdf

Sklenicka P., Zouhar J., Predicting the visual impact of onshore wind farms via landscape indices: A method for objectivizing planning and decision processes, *Applied Energy* 209, 445–454, 2018.

Stiller J., Rakowska A., Grabowski A., Kable i przewody (NN, SN, WN). *Oddziaływanie linii kablowych najwyższych napięć prądu przemiennego (AC) na środowisko* [Cables and wires (LV, MV, HV). Environmental impact of highest-voltage alternating current (AC) cable lines - in Polish], Europejski Instytut Miedzi, 2006.

Stryjecki M., Mielniczuk K., *Wytyczne w zakresie prognozowania oddziaływań na środowisko farm wiatrowych* [Guidelines for forecasting environmental impacts of wind farms - In Polish], Generalna Dyrekcja Ochrony Środowiska, 2011.

Sudra P., Bida-Wawryniuk Z., *Uwarunkowania planistyczno-prawne lokalizacji elektrowni wiatrowych w Polsce i w innych krajach europejskich* [Planning and legal conditions for the location of wind power plants in Poland and other European countries - in Polish], *Człowiek i Środowisko* 41(2), 2018.

Superuse, 2020. https://www.superusestudios.com/

SusWind._https://www.nccuk.com/what-we do/sustainability/https://ore.catapult.org.uk/, 2019.

Szasz R.Z., Leroyer A., Revstedt J., Numerical modelling of the ice throw from wind turbines, *The International Journal of Turbomachinery Propulsion and Power* 4(4), 10.3390, 2019.

Szczeciński S. Zespoły wirnikowe silników turbinowych [Turbine engine rotor assemblies - in Polish], WKiŁ, Warszawa, 1982.

Szuba M., *Linie i stacje elektroenergetyczne w środowisku człowieka* [Power lines and substations in the human environment - In Polish], Polskie Sieci Elektroenergetyczne S.A., Warszawa, 2008.

Szychowska M., Hafke-Dys H., Preis A., Kociński J., Kleka P., The influence of audio-visual interactions on the annoyance ratings for wind turbines, *Applied Acoustics* 129, 190–203, 2018.

Tammelin B., Cavaliere M., Holttinen H., Morgan C., Seifert H., Säntti K., *Wind energy production in cold climate (WECO)*, ETSU Contract Rep W/11/00452/REP, UK DTI, 1999.

The Working Group on Noise from Wind Turbines, *The assessment and rating of noise from wind farms*, Energy Technology Support Unit, ETSU-R-97, 1996.

Thornmann Recycling Sp. z o.o. https://thornmann.com.pl/#undefined

Tonin R., Brett J., Colagiuri B., The effect of infrasound and negative expectations to adverse pathological symptoms from wind farms, *Journal of Low Frequency Noise, Vibration and Active Control* 35(1), 77–90, 2016.

Uadiale S., Urbán E., Carvel R., Lange D., Rein G., *Overview of Problems and Solutions in Fire Protection Engineering of Wind Turbines*, Fire Safety Science, 2014.

UK Government Department for Communities and Local Government, National Planning Policy Framework, 2012.

Van den Berg G.P., Wind turbine power and sound in relation to atmospheric stability, *Wind Energy* 11, 151–69, 2008.

Van Kamp I., van den Berg F., Health effects related to wind turbine sound: an update, *International Journal of Environmental Research and Public Health* 18(17), 9133, 2021.

Vestas Online Business, Vestas Online Compact. Operator's Manual, Vestas Wind Systems A/S, 2019.

Vlaamse overheid - Departement Leefmilieu, Natuur en Energie, Toelichtingsnota nieuwe milieuvoorwaarden voor windturbines, 2012.

Watanabe T., Møller H., Low Frequency Hearing Thresholds in Pressure Field and in Free Field, *Journal of Low Frequency Noise, Vibration and Active Control* 9(3), 106–115, 1990.

Wegner S., Bareiss R., Guidati G., *Wind turbine noise*, Springer, Berlin, 1996.

Westerlund M., Social acceptance of wind energy in urban landscapes, *Technology Innovation Management Review* 10(9), 49–62, 2020.

Wind Energy. Code of good practice - in Polish. https://psew.pl/wp-content/uploads/2019/10/KDP-z-rekomendacjami-pa%C5%BAdziernik-2019.pdf

Wind turbine accident and incident compilation 2020. https://www.caithnesswindfarms.co.uk/

Wind Turbine Noise 2005, Berlin Niemcy, 2005. https://www.windturbinenoise.eu/content/conferences/3-wind-turbine-noise-2005/

Wind turbines. Fire protection guideline, CFRA Europe 22, Stockholm, 2010.

Wings for Living. https://wings-for-living.com/

Wolsink M., Planning of renewables schemes: deliberative and fair decision-making on landscape issues instead of reproachful accusations of non-cooperation, *Energy Policy* 35(5), 2692–2704, 2007.

World Bank Group, Environmental, Health and Safety Guidelines Wind Energy, 2015.

World Energy Transitions Outlook 2022: 1.5°C Pathway - Executive Summary (irena.org), 2022.

World Health Organization, *Environmental noise guidelines for the European region*, World Health Organization Regional Office for Europe, Copenhagen, 2018.

Yoon K., Gwak D.Y., Seong Y., Lee S., Hong J., Lee S., Effects of amplitude modulation on perception of wind turbine noise, *Journal of Mechanical Science and Technology* 30(10), 4503–4509, 2016.

Zagubień A., Analysis of acoustic pressure fluctuation around wind farms, *Polish Journal of Environmental Studies* 27(6), 2843–2849, 2018.

Zagubień A., *Pomiar tła akustycznego w środowisku - studium przypadków* [Measurement of background noise in the environment - a case study - in Polish], Rocznik Ochrona Środowiska, 20, 1498–1514, 2018.

Zagubień A., Ingielewicz R., The analysis of similarity of calculation results and local measurements of wind farm noise, *Measurement* 106, 211–220, 2017.

Zagubień A., Wolniewicz K., Everyday exposure to occupational/non-occupational infrasound noise in our life, *Archives of Acoustics* 41, 659–668, 2016.

Zagubień A., Wolniewicz K., *Domowe źródła hałasu niskoczęstotliwościowego* [Domestic sources of low-frequency noise - in Polish], *Rocznik Ochrona Środowiska* 19, 682–693, 2017.

Zagubień A., Wolniewicz K., *Domowe źródła hałasu niskoczęstotliwościowego* [Home sources of low frequency noise], *Rocznik Ochrona Środowiska*, 19, 682–693, 2017.

Zagubień A., Wolniewicz K., The impact of supporting tower on wind turbine noise emission, *Applied Acoustics* 155, 260–270, 2019.

Zagubień A., Wolniewicz K., The assessment of infrasound and low frequency noise impact on the results of learning in primary school - case study, *Archives of Acoustics* 45, 93–102, 2020.

Zajamšek B., Hansen K.L., Doolan C.J., Hansen C.H., Characterisation of wind farm infrasound and low-frequency noise, *Journal of Sound and Vibration* 370, 176–190, 2016.

Zajdler R., *Regulacje prawa krajowego dotyczące inwestycji w farmy wiatrowe (wybrane aspekty)* [Regulations of national law on investment in wind farms (selected aspects) - in Polish], Instytut Sobieskiego, Warszawa, 2012.

Zathey M., Studium przestrzennych uwarunkowań rozwoju energetyki wiatrowej *w województwie dolnośląskim* [Study of spatial determinants of wind power development in the Lower Silesian Province - In Polish], Wojewódzkie Biuro Urbanistyczne, Wrocław, 2010.

ZEBRA._https://www.lmwindpower.com/en/stories-and-press/stories/news-from-lm-places/zebra-project-launched, 2019.

Index

Note: **Bold** page numbers refer to tables and *italic* page numbers refer to figures.